TOXIC ALIPHATIC FLU

S0-DTY-997

ELSEVIER MONOGRAPHS

INDUSTRIAL TOXIC AGENTS

EDITED BY

ETHEL BROWNING, M.D.

FORMERLY H.M. MEDICAL INSPECTOR OF FACTORIES
MINISTRY OF LABOUR AND NATIONAL SERVICE
LONDON

ELSEVIER PUBLISHING COMPANY
AMSTERDAM LONDON NEW YORK PRINCETON

INDUSTRIAL TOXIC AGENTS

TOXIC ALIPHATIC FLUORINE COMPOUNDS

BY

F. L. M. PATTISON

M.A., PH.D., F.R.I.C., F.C.I.C.

PROFESSOR AND HEAD, DEPARTMENT OF CHEMISTRY
UNIVERSITY OF WESTERN ONTARIO
LONDON, ONTARIO, CANADA

WITH A FOREWORD BY

SIR RUDOLPH A. PETERS

M.D., F.R.S.

DEPARTMENT OF BIOCHEMISTRY
INSTITUTE OF ANIMAL PHYSIOLOGY
BABRAHAM HALL, CAMBRIDGE
LATE WHITLEY PROFESSOR OF BIOCHEMISTRY, OXFORD

ELSEVIER PUBLISHING COMPANY
AMSTERDAM LONDON NEW YORK PRINCETON
1959

SOLE DISTRIBUTORS FOR THE UNITED STATES OF NORTH AMERICA:

D. VAN NOSTRAND COMPANY, INC.

120 Alexander Street, Princeton, N.J. (Principal office)
257 Fourth Avenue, New York 10, N.Y.

SOLE DISTRIBUTORS FOR CANADA:

D. VAN NOSTRAND COMPANY (CANADA), LTD.

25 Hollinger Road, Toronto 16

SOLE DISTRIBUTORS FOR THE BRITISH COMMONWEALTH EXCLUDING CANADA:

D. VAN NOSTRAND COMPANY LTD.

358 Kensington High Street, London, W. 14

Library of Congress Catalog Card Number 59-12586

With 6 figures and 31 tables

ALL RIGHTS RESERVED
THIS BOOK OR ANY PART THEREOF MAY NOT BE REPRODUCED
IN ANY FORM (INCLUDING PHOTOSTATIC OR MICROFILM FORM)
WITHOUT WRITTEN PERMISSION FROM THE PUBLISHERS.

PRINTED IN THE NETHERLANDS BY
ZUID-NEDERLANDSCHE DRUKKERIJ N.V., 'S-HERTOGENBOSCH

615.95
P321

Contents

60/61-7

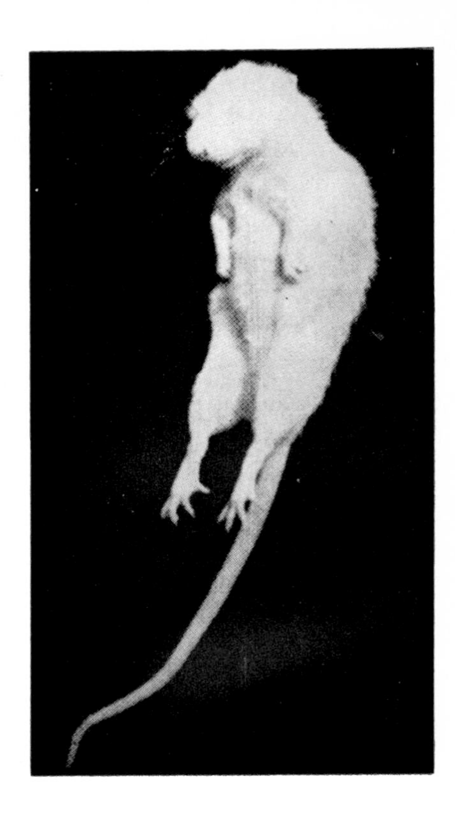

Rat in tetanic convulsion after poisoning by sodium fluoroacetate (Compound 1080).

(Reproduced with permission, R. A. PETERS, *Proc. Roy. Soc. (London)*, *B*, *139* (1952) 143.)

Foreword

In the expanded horizons of this age of space research, it is sometimes difficult to focus on terrestrial detail. Hence a book which reminds us of the far-reaching significance of small structural changes in chemistry is especially valuable. The field of aliphatic fluorine compounds has expanded rapidly during the last fifteen years. Though known earlier in Belgium and Germany, the war forced attention on fluoroacetate, which has developed with increasing interest during the last ten years. This has been due partly to the discovery that a well-known South African plant contains this poison, and partly to the realization of the very special properties conferred by fluorine, extending even to compounds of medical importance.

Since 1943, Professor Pattison has been engaged, firstly with Dr. B. C. Saunders in England and later in Canada with colleagues, in investigation of the organic chemistry and properties of different types of aliphatic fluorine compounds. His long experience has brought him into close touch with workers in biology, biochemistry and pharmacology. Any book by him in this field must therefore have a special authority. His primary object here is to present the collected facts which should be available quickly to those working with these compounds. Nevertheless with characteristic energy, Professor Pattison has given the book a much wider scope. It may be read with ease and profit by many others for its opening of new vistas, as well as for its revelation of the profound changes in biological properties induced by the introduction of one atom of fluorine into a molecule. I can commend it most warmly to the reader.

Cambridge
January, 1959

RUDOLPH A. PETERS

Preface

The purpose of this monograph is to provide a brief account of toxic aliphatic fluorine compounds. In a work of this size, it has not been possible to review in its entirety the relevant literature, which is now extensive. All that I can hope to produce, in fact, is a balanced, not an exhaustive, treatment of the subject. It is my intention to outline the chemistry, biochemistry, toxicology, pharmacology and uses of these compounds; to assemble the available information on their effects on man, with particular reference to diagnosis and treatment; to draw attention to certain criminological and forensic aspects of the subject; and finally, in reviewing the types of compounds which are toxic, to suggest rules for predicting the probable danger of a new or unknown fluoro compound.

It is perhaps appropriate at the outset to indicate the general coverage. A judicious selection of information relating to toxic compounds containing the C-F bond, published up to the end of 1958, comprises the major part of the book; also described are a few non-toxic compounds, the inclusion of which is necessary for the discussion. The following classes have *not* been included as being outside the scope of the monograph: various classes of non-toxic aliphatic fluorine compounds; aromatic fluorine compounds; certain fluorine compounds which owe their toxicity to moieties other than the C-F group, for example, sulphur pentafluoride, S_2F_{10} and compounds containing the P-F bond; inorganic fluorine compounds such as elementary fluorine, hydrogen fluoride, metal fluorides and interhalogen compounds; compounds toxic only by virtue of liberation of hydrogen fluoride, for example, acyl fluorides; and finally a large number of compounds which are probably toxic but for which no toxicity data have been presented (hence much early work, notably the pioneer researches of Swarts, has perforce been omitted). It will be clear from this list that fluoridation of drinking water, and the contentious arguments surrounding it, will not be discussed.

While many of the compounds described in this monograph cannot be considered as 'Industrial Toxic Agents' (the title of the series), it is possible nevertheless that they may be formed in small but dangerous amounts as byproducts in the rapidly increasing number of commercial processes involving fluorinated materials. Hence in the interest of safety, all those working with organic fluorine compounds should at least be aware of the types likely to be hazardous, even if such compounds may occur only as minor contaminants. In considering this point, it is appropriate to contrast toxicity and hazard in connection with poisons in general. Thus, a highly toxic compound might be non-volatile or unstable or possess a pronounced and obvious odour, and hence pose less of a hazard than a less toxic but more insidious material. By the same token, while there are several more toxic substances known, few are as hazardous as the fluoroacetates and related compounds.

The chemistry is included throughout not so much to indicate preparations and properties *per se*, as to provide a means of assessing which toxic compounds might accidentally be formed in any particular process. Boiling points are tabulated as an indication of volatility, which in turn is a rough guide to the quantity of material likely to be encountered in the atmosphere surrounding the point of generation. Other physical constants, such as refractive indices, were considered to be superfluous in a general survey of this kind; they may be found in the cited literature.

The paucity of information on the effects of toxic fluorine compounds on man is unfortunate (but not altogether surprising); such data as are available have arisen mainly through accidental intoxication. The majority of toxicity figures have been obtained using *mice*. It must be emphasized that, although such results provide the basis for a semi-quantitative comparison between different compounds, they cannot be extended to apply to man; all that may be conceded is that the most toxic compounds to mice are probably the most toxic compounds to man. Such a grudging concession is justified by results and inferences drawn from a large body of data obtained with other experimental animals.

It is possible, or even probable, that some of the biochemical conclusions (particularly those outlined in Chapter 4) may require rejection or, at least, radical alteration as more results accrue; in short, the tentative nature of some of the conclusions cannot be stressed too strongly. But however inadequate the present work may prove to be, perhaps it may provide an incentive for further research—even if this is motivated merely by a desire to right some of the more flagrant errors. In this connection, may I request readers to assist in improving possible future editions of this book by sending to me corrections and criticisms, however frank?

In the course of reviewing the literature and preparing the manuscript of this book, I was frequently reminded of my early experiences in fluorine chemistry while working as a graduate student under the direction of Dr. B. C. Saunders in Cambridge; I should like to record my gratitude and appreciation for such a stimulating start in such a fascinating field of research.

I wish also to express my deepest appreciation to the many persons who came to my aid in answering questions, supplying information, and proffering constructive criticism and advice. I am particularly grateful to Sir Rudolph Peters, Cambridge, Dr. M. B. Chenoweth, Midland, Michigan, Dr. W. C. Howell, London, Ontario, and Mr. R. W. White, London, Ontario, for so kindly reading the entire manuscript and for tendering so many pertinent comments. My sincere thanks are due also to Dr. M. B. Chenoweth for providing the section on the treatment of fluoroacetate poisoning, and to Dr. E. H. Bensley, Montreal General Hospital, for advice in preparing the hospital case histories in a palatable form. To my wife Anne I express my warm gratitude for her constant support and encouragement.

I am greatly indebted to the Chief Librarian and staff of King's

College Library, Newcastle upon Tyne, for their hospitality and help during part of the summer of 1958; to Mrs. Verna Leah for expert workmanship in typing the manuscript; and finally to Dr. Ethel Browning for her unfailing help and courtesy.

London, Ontario
May, 1959

F. L. M. PATTISON

Acknowledgements

The author's thanks are due to the following for permission to reproduce figures: Sir Rudolph Peters and the Council of the Royal Society for the Frontispiece; the Editor of *Farming in South Africa* for the two pictures contained in Fig. 3; the Editor of the *Journal of Ecology* for Fig. 4; and Sir Rudolph Peters and the Editor of *Endeavour* for Fig. 6.

1

Introduction and general survey

'Mordre wol out, that see we day by day'.
CHAUCER, *The Nonnes Preestes Tale*, 1. 232

'In poison there is physic'.
SHAKESPEARE, *II Henry IV*, Act I, sc. 1, 1. 137.

Criminologists are continually on the alert for new types of poisons which might be used in irresponsible hands for committing 'the perfect murder'. Some of the requirements for such a poison include the following. (a) It should possess high toxicity. (b) It should be inconspicuous, particularly in regard to odour and taste. (c) It should be sufficiently stable to survive rather drastic conditions, such as suspension in boiling coffee. (d) It should operate with a delayed action, in order that the murderer may cover his tracks before death occurs. (e) The poison should induce no obvious pathological changes which could be spotted on post-mortem examination. (f) There should be no reliable means of analysing for residual traces of the toxic agent in the body of the victim. (g) There should be no known antidote for saving the life of the victim after toxic symptoms have developed.

From a study of this monograph, it will be seen that, until recently, fluoroacetic acid and many other monofluoro compounds satisfied these criteria to an alarming degree; so much so that several of the compounds were examined during the last war for possible use as chemical warfare agents. Modern research has now provided reliable methods for the detection of these poisons; hence the danger of their misuse has diminished greatly. It is nevertheless of great importance that forensic workers of all sorts should be more aware of these very dangerous compounds, referred to, with justification, as being 'in the category of the most poisonous substances known'[3]. It is the hope of the author that a more widespread knowledge of these compounds may act as a deterrent to their

References p. 10

unlawful use, and that interest may be stimulated in the search for more satisfactory methods of medical and veterinary treatment.

For obvious reasons, there is a dearth of information available covering the effects of these compounds on man. However, several workers have felt justified in drawing inferences from the plethora of data obtained using experimental animals. Thus, although the primary concern of the monographs in this series is the effect of toxic agents on man, results of animal experiments germane to this restricted biological field have therefore been freely included. It will be appreciated however that the contents do not constitute an exhaustive review of results for all species of animals.

In any survey* of the development of toxic fluorine compounds, pride of place must go to fluoroacetic acid, FCH_2COOH and its simple derivatives, collectively known as 'fluoroacetates'. The historical background of this acid is both interesting and unusual. It was first prepared by Swarts[18] in 1896, but no mention was made of its toxicity. Some forty-five years later, Polish chemists[4], escaping to England early in the war, reported on the poisonous effects of its methyl ester. This led to an extensive study of fluoroacetic acid and its derivatives in England[8] and the United States[5], from which emerged the fact that fluoroacetates are powerful poisons, being more toxic than gases such as hydrogen cyanide, phosgene and carbon monoxide. Then two or three years later, in 1944, came the astonishing announcement by Marais[7] that fluoroacetic acid is the toxic principle of the very poisonous South African plant *Dichapetalum cymosum*, a well-recognized hazard to cattle. It thus follows that African natives, in using extracts of this and related plants as arrow poisons, had anticipated by many generations the proposed use of fluoroacetates as chemical warfare agents. During the decade that followed this announcement, the biochemistry and pharmacology of fluoroacetates were investigated extensively by Peters[16] in England and Chenoweth[2] in the United States. It has now been well established that fluoroacetic acid is converted

* Many of the points mentioned in this survey will be discussed more fully in subsequent chapters, and a more complete bibliography will be presented.

TABLE I

TOXICITY OF FLUOROACETATE†

Species	*LD_{50}, mg/kg*	*Route*††
Dog	0.06	I.V.
Cat	0.2	I.V.
Sheep	0.3°	O.
Guinea pig	0.35	I.P.
Rabbit	0.5	I.V.
Horse	1.0	O.
Man	2–10°°	O.
Rat	5	I.P. or S.C.
Mouse	7	I.P.
Frog	150	S.C.
South African clawed toad	> 500	I.P.

† Compiled by Chenoweth[2].
†† Routes: I.V., intravenous; I.P., intraperitoneal; O., oral; S.C., subcutaneous.
° Reported by Peters[16] for English sheep.
°° The figure for man should be accepted with considerable reserve.

in vivo to fluorocitric acid, which in turn blocks the tricarboxylic acid cycle by inhibiting the enzyme aconitase[16]; the resultant deprivation of energy is thought to result in impairment of cellular function, destruction of permeability barriers and, ultimately, death. The researches leading up to this conclusion 'have paid dividends for pure biochemistry: the revelation of new inhibitors, better light on an obscure piece of biochemistry, and much needed information upon the bridge relating biochemistry to physiology'[16].

For expressing toxicity, the term LD_{50}* is most commonly used. It can be seen from the few examples presented in Table I that the value for fluoroacetic acid (or its salts or simple esters) varies extraordinarily widely from one species of animal to another.

* The term LD_{50} signifies the dose, expressed in milligrams of compound per kilogram body weight of animal, required to kill 50% of a batch of animals. A low value therefore represents high toxicity. The majority of compounds were administered in propylene glycol as solvent.

References p. 10

For example, the figures for the dog and the South African clawed toad differ by a factor close to 10,000. Since the mouse was the most commonly used experimental animal, the figure of 7 mg/kg was used as a standard of comparison for all new compounds prepared. The figure for the rat emphasizes the effectiveness of Compound 1080 as a rodenticide, but the greater toxicity of fluoroacetates to dogs and cats points a warning to animal lovers. The figure for man, which is at best rather speculative, suggests that the lethal dose for an 'average' individual of 70 kg might be about five drops; that man is relatively less sensitive than some species to fluoroacetate was shown by Adrian, who drank a dose sufficient to produce a urine toxic to guinea pigs[15]; medical aspects are treated in Chapter 2 (p. 47). It has been reported[4] that a horse died after drinking 10 litres of water containing a few drops of fluoroacetate, and that a dog died after eating some of the meat from this horse. Insects are apparently very sensitive to fluoroacetate; for example, fleas are killed by feeding on poisoned rats[6]. There is marked variation in the symptoms of poisoning between different species of animals; this point is developed in Chapter 2.

During the early investigation of the fluoroacetates, 2-fluoroethanol, FCH_2CH_2OH and some simple derivatives were prepared and examined[8, 17]. The parent compound had a toxicity almost identical with that of fluoroacetic acid, suggesting that oxidation had occurred *in vivo*. Of its derivatives, those that could readily be metabolized to 2-fluoroethanol were toxic, whereas the others were not.

The next development centered around the homologous series of ω-fluorocarboxylic esters, $F(CH_2)_nCOOR$. It was soon established[1] that a remarkable alternation in toxicity occurred with the ascent of the series: thus, if the total number of carbon atoms in the acid moiety is even, the compound is toxic and produces symptoms in animals similar to those produced by fluoroacetic acid, whereas if the total number is odd, the compound is non-toxic. For example, ethyl 6-fluorohexanoate, $F(CH_2)_5COOC_2H_5$ is toxic, but ethyl 7-fluoroheptanoate, $F(CH_2)_6COOC_2H_5$ is not. More recently, the free acids have been prepared[13], the results of which are listed in Table II and represented diagrammatically in Fig. 1. There is of course considerable variation between members in the same toxicity 'group', but the overall trend is unmistakable. On a molar basis, 8-fluoro-octanoic acid, $F(CH_2)_7COOH$ is twenty-three times more toxic than fluoroacetic acid.

TABLE II

TOXICITY OF ω-FLUOROCARBOXYLIC ACIDS, $F(CH_2)_nCOOH$

Formula of acid	*LD_{50} for mice (intraperitoneal) mg/kg*
FCH_2COOH	6.6
$F(CH_2)_2COOH$	60
$F(CH_2)_3COOH$	0.65*
$F(CH_2)_4COOH$	> 100
$F(CH_2)_5COOH$	1.35
$F(CH_2)_6COOH$	40
$F(CH_2)_7COOH$	0.64
$F(CH_2)_8COOH$	> 100
$F(CH_2)_9COOH$	1.5*
$F(CH_2)_{10}COOH$	57.5
$F(CH_2)_{11}COOH$	1.25
$F(CH_2)_{17}COOH$	5.7

* Sodium salt used.

This pronounced alternation in toxicity has been correlated with the β-oxidation theory of fatty acid metabolism[14]. It can readily be seen that the toxic acids can in all cases be degraded to the toxic fluoroacetic acid, whereas the non-toxic acids can be oxidized only so far as the non-toxic 3-fluoropropionic acid (or its non-toxic catabolites). The increase in toxicity of the higher members may be associated with their greater lipid solubility.

It thus became apparent from this and other early work that *any compound which can form fluoroacetic acid by some simple biochemical process is toxic and gives rise to symptoms in animals similar to those produced by fluoroacetic acid itself*[1, 8]. This important generalization forms the basis for much later work.

References p. 10

Because of the interesting toxicological results associated with the ω-fluorocarboxylic acids, a wide variety of closely related compounds was prepared and examined[9–11]. It was anticipated that enhancement of toxicity might occur with the introduction

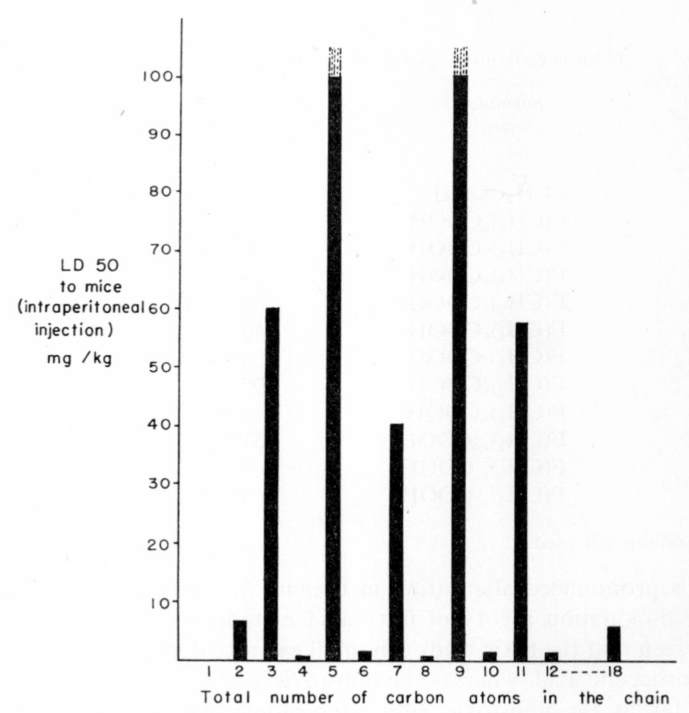

Fig. 1. Toxicity of ω-fluorocarboxylic acids, $F(CH_2)_nCOOH$.

of such groupings as -CN and -SCN in place of the original -COOH grouping. In each instance, an attempt was made to correlate the toxicity pattern with known metabolic detoxication processes; in all cases, the results tallied with expectations. Three typical series are mentioned below; fuller treatment is given in Chapter 4. The results of the work made it possible to draw up simple rules to act as a rough guide in predicting if a new or unknown fluorine compound might be toxic or not (see pp. 158, 196).

Since the biological oxidation of alcohols to acids is well known, the homologous series of ω-fluoroalcohols, $F(CH_2)_nOH$[12] was first investigated (p. 127). An alternation in toxicity strikingly similar to that found for the ω-fluorocarboxylic acids was established, with the same generalizations regarding the odd and even members (Table III and Fig. 2). This fact therefore provides confirmation of the conversion of alcohols to acids *in vivo*. Like 8-fluoro-octanoic acid, 8-fluoro-octanol, $F(CH_2)_8OH$ is twenty-three times more toxic than fluoroacetic acid on a molar basis.

TABLE III

TOXICITY OF ω-FLUOROALCOHOLS, $F(CH_2)_nOH$

Formula of alcohol	*LD_{50} for mice (intraperitoneal) mg/kg*
FCH_2CH_2OH	10
$F(CH_2)_3OH$	46.5
$F(CH_2)_4OH$	0.9
$F(CH_2)_5OH$	> 100
$F(CH_2)_6OH$	1.24
$F(CH_2)_7OH$	80
$F(CH_2)_8OH$	0.6
$F(CH_2)_9OH$	32
$F(CH_2)_{10}OH$	1.0
$F(CH_2)_{11}OH$	> 100
$F(CH_2)_{12}OH$	1.5
$F(CH_2)_{18}OH$	4.0

Aliphatic amines are known to be converted *in vivo* first to the corresponding aldehydes and then to the acids ($RCH_2NH_2 \rightarrow RCHO \rightarrow RCOOH$). Hence it was expected that the ω-fluoroalkylamines $F(CH_2)_nNH_2$ would produce this same alternation in toxicity, with the even* members being the more toxic. The results shown in Table XXV (p. 148) bear this out.

Aliphatic nitriles have been reported to be degraded *in vivo* to hydrogen cyanide and the next lower acid ($RCH_2CN \rightarrow RCOOH + HCN$). Consequently, the ω-fluoronitriles $F(CH_2)_nCN$ containing an *odd** number of carbon atoms were expected to be the toxic

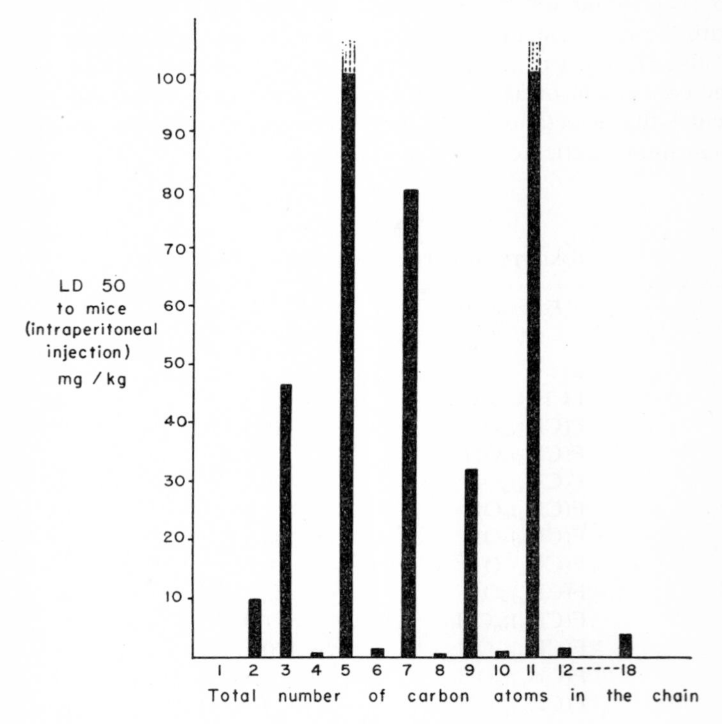

*Fig. 2**. Toxicity of ω-fluoroalcohols, $F(CH_2)_nOH$.*

* The terms 'even' and 'odd' refer to the *total* number of C atoms in the chain.

** Bar-graph representation as in Fig. 1 and 2 demonstrates more vividly the toxicity pattern, but is less accurate than tabular presentation of figures. Tables alone will be used from this point.

members. The figures presented in Table XXIII (p. 144) provide confirmation for this suggestion.

From the results of this and other work emerged[11] a new and general procedure for elucidating the biochemical breakdown *in vivo* of aliphatic compounds, utilizing the ω-fluorine atom, with its characteristic toxicological properties, as a 'tag'. The arguments and conclusions are based on crude toxicity figures, and the rationale resides in the characteristic toxicological behaviour of the ω-fluorocarboxylic acids. The method, based on results given in Chapter 4, is described in Chapter 5 (p. 182).

It should be added at this point that, while most of the compounds included in this book contain only one fluorine atom (and this in the ω-position), some polyfluoro compounds, notably certain polyfluoroalkenes, are also very toxic and should be handled with extreme caution (see p. 124).

Uses and potential applications of some pharmacologically-active aliphatic fluorine compounds are outlined in Chapter 5. Two of these have advanced beyond the exploratory stage: knowledge of sodium fluoroacetate as a pest exterminator is extensive, and clinical trials with the general anaesthetic Fluothane may be considered as being exceptionally promising. Other uses, most of which are still tentative, have been included to indicate trends and to offer clues for further work.

None of the compounds so far examined has shown any pronounced increase in toxicity over that of the corresponding member of the ω-fluorocarboxylic acid series. This is hardly surprising, because, as was not known at the time, the antepenultimate toxic agent in all cases is the same, namely fluoroacetic acid*, which in turn is converted *in vivo* to fluoroacetyl-coenzyme A and then to fluorocitric acid[16]. Hence it is reasonable to conclude that, in the absence of information to the contrary, treatment of poisoning caused by any of the ω-fluoro compounds should be essentially the same as that of poisoning caused by fluoroacetic acid itself.

* It is likely that in many instances fluoroacetyl-coenzyme A may be formed directly without the intermediate formation of fluoroacetic acid itself.

The reasons outlined in the above survey justify the full treatment accorded to fluoroacetic acid and its simple derivatives in the next chapter.

REFERENCES

1. Buckle, F. J., Pattison, F. L. M., and Saunders, B.C. (1949) Toxic fluorine compounds containing the C-F link. Part VI. ω-Fluorocarboxylic acids and derivatives. *J. Chem. Soc., 1949:* 1471.
2. Chenoweth, M.B. (1949) Monofluoroacetic acid and related compounds. *J. Pharmacol. Exptl. Therap., II, 97:* 383; *Pharmacol. Revs., 1:* 383.
3. Foss, G. L. (1948) The toxicology and pharmacology of methyl fluoroacetate (MFA) in animals, with some notes on experimental therapy. *Brit. J. Pharmacol., 3:* 118.
4. Gryszkiewicz-Trochimowski, E., Sporzynski, A., and Wnuk, J. (1947) Recherches sur les composés organiques fluorés dans la série aliphatique. II. Sur les dérivés des acides mono-, di- et tri-fluoroacétiques. *Rec. trav. chim., 66:* 419.
5. Kharasch, M. S. (1943–1945) Unpublished results.
6. Macchiavello, A. (1946) Plague control with DDT and '1080'. Results achieved in a plague epidemic at Tumbes, Peru, 1945. *Am. J. Public Health, 36:* 842.
7. Marais, J. S. C. (1944) Monofluoroacetic acid, the toxic principle of 'Gifblaar', *Dichapetalum cymosum* (Hook) Engl. *Onderstepoort J. Vet. Sci. Animal Ind., 20:* 67.
8. McCombie, H., and Saunders, B. C. (1946) Fluoroacetates and allied compounds. *Nature, 158:* 382.
9. Pattison, F. L. M. (1953) Toxic fluorine compounds. I. *Nature, 172:* 1139.
10. Pattison, F. L. M. (1954) Toxic fluorine compounds. II. *Nature, 174:* 737.
11. Pattison, F. L. M. (1957) From war to peace: toxic aliphatic fluorine compounds. *Chem. in Can., 9, No. 8:* 27.
12. Pattison, F. L. M., Howell, W. C., McNamara, A. J., Schneider, J. C., and Walker, J. F. (1956) Toxic fluorine compounds. III. ω-Fluoroalcohols. *J. Org. Chem., 21:* 739.
13. Pattison, F. L. M., Hunt, S. B. D., and Stothers, J. B. (1956) Toxic fluorine compounds. IX. ω-Fluorocarboxylic esters and acids. *J. Org. Chem., 21:* 883.

14. PATTISON, F. L. M., and SAUNDERS, B. C. (1949) Toxic fluorine compounds containing the C-F link. Part VII. Evidence for the β-oxidation of ω-fluorocarboxylic acids *in vivo*. *J. Chem. Soc., 1949:* 2745.
15. PETERS, R. A. (1952) Lethal synthesis. [Croonian lecture] *Proc. Roy. Soc. (London), B 139:* 143.
16. PETERS, R. A. (1957) Mechanism of the toxicity of the active constituent of *Dichapetalum cymosum* and related compounds. *Advances in Enzymology and Related Subjects of Biochemistry*, Vol. XVIII, Interscience Publishers, Inc., New York, p. 113.
17. SAUNDERS, B. C., STACEY, G. J., and WILDING, I. G. E. (1949) Toxic fluorine compounds containing the C-F link. Part II. 2-Fluoroethanol and its derivatives. *J. Chem. Soc., 1949:* 773.
18. SWARTS, F. (1896) Sur l'acide fluoroacétique. *Bull. Acad. roy. Belg.*, [3], *31:* 675.

2

Fluoroacetates

A glance at the number of entries in *Chemical Abstracts* listed under the heading of 'acetic acid, -fluoro' gives some indication of the surge of interest in fluoroacetates which has occurred since the end of the last war. For example, in 1944 there are no entries; in 1948 the entries occupy two inches vertically; in 1950, three inches; and in 1954, five and three quarter inches. Various reasons may be put forward to account for this increase, among which may be cited the following important developments: (a) the announcement by Marais[100] that fluoroacetic acid (or some simple derivative) occurs as a toxic constituent of at least one plant species; (b) the introduction in the United States of sodium fluoroacetate as a rodenticide and general mammalian pest control agent[82]; and (c) the release by the appropriate government agencies of the results of a large volume of secret war-time research.

The contents of this chapter relate to fluoroacetic acid and its simple derivatives and analogues; the latter are restricted to those compounds containing two carbon atoms in the fluorine-containing moiety. All long-chain ω-fluoro compounds and compounds which, although possessing two carbon atoms, are related only remotely to fluoroacetic acid are relegated to subsequent chapters. In illustration of this arbitrary division, 2-fluoroethanol, FCH_2CH_2OH, fluoroacetaldehyde, FCH_2CHO and 2-fluoroethyl fluoroacetate, $FCH_2COOCH_2CH_2F$ are included, whereas 2-fluoroethyl isocyanate, FCH_2CH_2NCO and 4-fluorobutanol, $F(CH_2)_4OH$ are discussed later. While it is hoped that this chapter provides a reasonably balanced account of the preparation and

Fig. 3. Dichapetalum cymosum (Gifblaar):

(a) a specimen of the plant growing on the veld;
(b) the fruit and leaves of the plant.

(Reproduced with permission, D. G. STEYN, *Farming in S. Africa, 14* (1939) 285.)

properties of the fluoroacetates, space limitations do not allow of an exhaustive survey of the copious literature on the subject.

OCCURRENCE

A poisonous plant known as 'gifblaar' *(Dichapetalum cymosum)* occurring in South Africa has long been known as one of the most deadly stock poisons (Figs. 3 and 4). Sir Arnold Theiler[168] in 1902 first confirmed its toxic nature by feeding the plant to oxen and rabbits with fatal results. It was recognized as a deadly poison by the early Voortrekkers who entered the Transvaal; indeed it was they who named it 'gifblaar' (*i.e.* poison leaf). It is an underground shrub possessing branches, many of which attain great length and from which small shoots ascend to the surface, there giving rise to the tufts of green lanceolate-ovate leaves and inconspicuous flowers so painfully familiar to the local farmers. Because of the

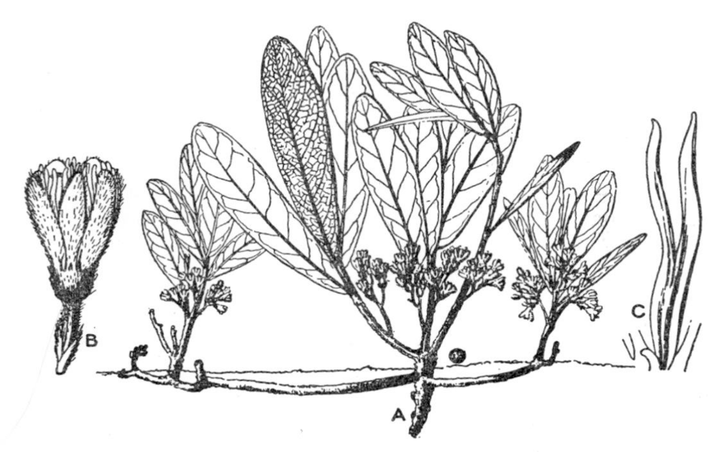

Fig. 4. Dichapetalum cymosum (Gifblaar).
A, the plant, showing habit and underground stems (reduced)
B, the flower (enlarged)
C, bifid petal with scale at the base (enlarged)

(Reproduced with permission, J. BURTT DAVY, *J. Ecol., 10* (1922) 211.)

very great depth to which the underground stems can penetrate (sixty feet or more[164]), the plant sprouts early in the summer before the first rains have fallen and when the veld is still dry; the resultant green patches of gifblaar are thus very attractive to cattle. It is toxic only during the spring (August to November) and autumn (March to May), and the younger the leaves, the greater their virulence[5, 161] (about three-quarters of an ounce of the young leaves is lethal to a goat). This variation in toxicity is of great importance from a stockowner's point of view, since it is just at these two seasons that fresh green grazing is so scarce[144]. The onset of symptoms is apparently accelerated when the animal partakes of water immediately after feeding on gifblaar. Steyn has provided some interesting and colourful records of the effects and attempted prevention of gifblaar poisoning[161, 163, 164]. Some idea of its action on cattle may be obtained from the following account by Theiler, quoted by Steyn[168]:

'In cattle, symptoms may appear within twelve hours after ingestion of the plant. The animal lies down with the head backwards, and when forced to rise it staggers, lifting the feet abnormally high, urinates very often, and immediately lies down again. When standing, the front legs are held well forward and the hind legs tucked under the body. The heart action is increased, the pulse is very soft and hardly perceptible. The respirations are increased, shallow in nature and the animals sometimes moan on expiration. Water and food are refused and rumination is absent. The animal shows great uneasiness, getting up and lying down and moving from side to side. Pronounced symptoms of hyperaesthesia are present. Furthermore there is a quivering of the muscles, especially those of the shoulder, and all the reflexes are exaggerated. The vision of the animal is impaired, and if allowed to walk it does not avoid objects but walks straight up against them. Salivation is increased and the animal very often grinds the teeth and therefore a frothy discharge comes from the mouth. After some time the animal gets very dull, the ears are drooping, and it is unable to rise. Diarrhoea seldom occurs. Death may occur as soon as twelve hours after the first symptoms have appeared.'

No particularly satisfactory methods of treatment have been reported, other than denying animals access to water for about two days, administering a purgative[163] at once, and dosing with strong, black coffee every few hours. Steyn records[164] that some farmers favour the administration of vinegar and kaffir beer. This recommendation is very interesting when viewed in the light of much later work involving the use of acetic acid derivatives and of ethanol in the treatment of fluoroacetate poisoning (p. 37). The farmers at this time of course had no notion of the nature of the toxic principle, and their recommendation thus provides another instance of the extraordinary luck or intuition involved in the empirical discoveries of herbal and other 'household' remedies.

Since treatment is largely ineffective, the only recommendation for safeguarding cattle is eradication of the plant. Total eradication is impracticable because of the extensive underground stem system; the commonest procedures involve erecting fences around infested areas and hand picking using native labour. Gifblaar-infested veld can however be used safely in the summer after good rains have fallen and when grazing is abundant. It may also be used in winter since the plant is killed by frost unless it grows in protected spots. It is advisable when danger of poisoning is present to allow the cattle to drink only in the morning, since, as mentioned above, large quantities of water apparently aggravate the symptoms.

All the above work on the ecology, toxicology and control of gifblaar proceeded without any knowledge of the active principle. Although studied earlier[144], it was not until 1943–4 that Marais in a very important investigation succeeded in isolating it in the form of its potassium salt[99]; this was subsequently proved to be *potassium fluoroacetate*[100]. Until this time, no naturally-occurring organic fluorine compounds had been known; indeed, this announcement by Marais completes the list of naturally-occurring organic halogen compounds, since the other three halogens all occur in organic compounds in nature. The biosynthesis of fluoroacetate *in situ* presents several interesting problems[31], not the least being the mechanism by which the C-F link is formed; and Saunders[148] has drawn attention to the related problem of the inferred breakdown

References p. 68

of fluoroacetate in the soil. Since a fundamental characteristic of all enzymes is the ability to promote *reversible* catalysis, it is not improbable that a single enzyme, as yet unrecognized, may be responsible for both effects; it is relevant to recall at this point the one recorded case of enzymic cleavage of the C-F bond[76]. There is some evidence that the soil in which gifblaar thrives is rich in fluoride[148].

At least one other naturally-occurring toxic compound containing the C-F link has been investigated. This is discussed in Chapter 3 (p. 83).

HISTORICAL DEVELOPMENT

It is unfortunate that much of the work on the fluoroacetates was carried out during the war in belligerent countries. The resultant delays in publication make the task of assessing and allocating priorities very hard. The account which follows has been prepared by taking the published literature at its face value; this seemed to be the only possible valid approach.

Fluoroacetic acid, FCH_2COOH was first prepared by Swarts[166] in Belgium in 1896. Some of its physical properties were recorded, but no mention was made of its toxicity. Swarts was also responsible for the first preparation of 2-fluoroethanol in 1914[167], but again toxicity was not mentioned. In 1930 fluoroacetic acid was patented as a mothproofing agent[169].

It was not until some forty years after Swarts' original preparation that Gryszkiewicz-Trochimowski and colleagues, working at the Warsaw Polytechnic during the period 1935–1939, carried out extensive researches[59, 60, 62, 63] into the preparation and properties of the fluoroacetates. Owing to exigencies of war, many records were lost and the work was not published until 1947. Their original discovery of the toxic effects of fluoroacetate stemmed from work with iodoacetates and their lacrimatory properties. When a bomb containing methyl iodoacetate explodes, a purple cloud results (due to liberation of free iodine), which in turn gives easy warning to field troops. In an examination of related compounds which would not possess this undesirable property, methyl fluoroacetate was prepared (by Swarts' original method) and assessed for lacrimatory action by being applied to the eye of a rabbit. No lacrimation was observed, but the rabbit died. This naturally led to a systematic study of fluoroacetates. A general method of preparation was elaborated[62] involving the reaction of aliphatic halogen compounds with finely powdered anhydrous potassium fluoride in an autoclave at elevated temperatures (150–250 °); necessary conditions included rigorous exclusion of moisture and vigorous agitation, and yields varied between 20 % and 90 %. Using this procedure for the introduction of the fluorine atom, the authors prepared many fluoroacetates and related compounds, including methyl fluoroacetate, 2-fluoroethyl fluoroacetate, methyl 2-fluoropropionate, methyl fluoroformate, epifluorohydrin, methyl fluoromalonate, 2-fluoroethanol, 3-fluoropropanol, 4-fluorobutanol, 3-fluoropropionic acid and 4-fluorobutyric acid.

In the course of the ensuing pharmacological examination, it became apparent[63] that the fluoroacetates were extremely toxic to a wide variety of animals by all the common routes of administration (p. 34). The symptoms of intoxication were considered to occur in cycles, each cycle involving three phases: (a) lazy and apathetic (30 to 120 minutes); (b) excitable, inability to see obstacles, enraged (about 30 min.); and (c) convulsions (2 to 4 min.). After 2 to 4 such cycles, the animals died of asphyxiation with convulsive symptoms. No post-mortem lesions could be observed. Of all the animals tested, only monkeys were considered to be strangely resistant to methyl fluoroacetate; even after massive doses (for example 2000 mg subcutaneously) they showed only the symptoms of the first phase much prolonged, after which they recovered completely. The authors, experimenting on themselves, observed that man can tolerate a rather large quantity of the poison: for example, they survived a concentration in the gas chamber nearly fifty times as great as that which killed dogs and other animals.

From a consideration of the many compounds examined, the authors concluded that the FCH_2COO^- anion was responsible for toxicity. Any compound which could form it by hydrolysis or oxidation was toxic, whereas those which could not were non-toxic;

References p. 68

examples of the latter included ω-fluoroacetophenone, methyl 2-fluoropropionate, 2-fluoroethyl bromide, difluoroacetic acid and trifluoroacetic acid, none of which could give rise to the FCH_2COO^- anion *in vivo*.

The only other work on toxic fluorine compounds reported during the pre-war period was that of Schrader in Germany[156–158]. Possibly because of the vagaries of state security and the natural reticence associated with industrial discoveries, little of this appeared in the open literature at that time. Although started in 1934, it came to light only after the war through the agency of teams of investigators working for the British Intelligence. The primary purpose of the work was to develop new types of insecticides. The first compounds to be prepared and examined were acid fluorides (RCOF and RSO_2F), of which methanesulphonyl fluoride, CH_3SO_2F was outstanding as a fumigant against grain weevils and aphids. This led to the investigation of other sulphonyl fluorides, RSO_2F, and of various related classes of compounds, including carbamyl fluorides, R_2NCOF, urethanes, $FCH_2CH_2OCONHR$, alkyl and aryl fluorosulphonates, FSO_2OR, fluorosulphonamides, FSO_2NR_2 and 2-fluoroethylamines, $FCH_2CH_2NR_2$.

In 1935, Schrader prepared 2-fluoroethanol and 2-fluoro-2′-hydroxydiethyl ether, $FCH_2CH_2OCH_2CH_2OH$, both of which were shown to be toxic to warm-blooded animals; 2-fluoroethanol was patented as a rodenticide. Many esters, urethanes and ethers of 2-fluoroethanol were then examined, of which bis-2-fluoroethyl sulphite was particularly effective against suctorial insects, yet showing no phytotoxicity. Some of the urethanes were found to be useful rodenticides.

But the most interesting and far-reaching achievement of Schrader in the field of aliphatic fluorine chemistry was the discovery of the first known systemic insecticides. This phase of the work is described on p. 173, including an account of the first practical use of this type of insecticide. The most active of these new compounds was the methylal derived from 2-fluoro-2′-hydroxydiethyl ether, $CH_2(OCH_2CH_2OCH_2CH_2F)_2$.

Final mention in this brief review of Schrader's work must be given to fluoroacetic acid itself, which was shown to be about as toxic to warmblooded animals as 2-fluoroethanol. The use of the salts of fluoroacetic acid as rodenticides was patented before the war.

Such then was the historical background of the fluoroacetates in the pre-war period up to 1939, the major advances having been made in Belgium, Poland and Germany. The next development occurred early in the war, when Sporzynski[62, 63], one of Gryszkiewicz-Trochimowski's colleagues, escaped to England from Poland and directed the attention of British chemists to the compound methyl fluoroacetate, which, as described above, had been shown to be outstandingly toxic. Under the direction of McCombie and Saunders, a team of chemists and physiologists in Cambridge accordingly undertook the systematic study of fluoroacetates and related compounds as potential chemical warfare agents[105]. A large variety of new compounds was synthesized and examined[22, 149, 150, 152, 153], and new preparative techniques were developed. From the toxicological results obtained, important relationships between physiological action and chemical constitution emerged, thereby providing independent confirmation for the generalization of Gryszkiewicz-Trochimowski outlined above. Methyl fluoroacetate was prepared from methyl chloroacetate and potassium fluoride by heating the reactants in an inclined, rotating autoclave for 4 hours at 220° (conversion, 54%; yield, 60%). It was found to be a powerful convulsant poison with a delayed action. The toxicities of many other simple derivatives of fluoroacetic acid were similar to that of methyl fluoroacetate. However, when one or both of the non-acidic hydrogens in the acid moiety were replaced by other atoms or groups, the compounds showed negligible toxicity. 2-Fluoroethanol, prepared in essentially the same manner as methyl fluoroacetate, was about equally toxic. The brief list that follows contains some representative compounds. Thus there emerges the inescapable conclusion that those compounds which can form fluoroacetic acid by hydrolysis and/or oxidation are toxic. Further reference to the work of the Cambridge team will be made in the sections that follow, but it is appropriate at this point to draw attention to the recent book by Saunders[148]

References p. 68

Toxic	*Non-toxic*
FCH_2COOH	$F_2CHCOOH$
FCH_2COOR	F_3CCOOH
FCH_2CONH_2	$CH_3CHFCOOR$
	$(CH_3)_2CFCOOR$
	$FCHClCOOR$
	$FCCl_2COOR$
	$FCOOR$
FCH_2CHO	
FCH_2CH_2OH	FCH_2CH_2Cl
FCH_2CH_2OCOR	
FCH_2COCl	$ClCH_2COF$

in which a detailed account is given of the extensive research programme which he developed and directed.

While the research in Cambridge was proceeding, Kharasch and colleagues in Chicago were working on related problems. Reports were freely exchanged between the two laboratories. Unfortunately the American work has not yet been published. However, some idea of the extensive researches may be obtained from an examination of the article by Redemann *et al.*[143] which records some physical properties of thirty fluoro compounds supplied by Kharasch.

Towards the end of the war, two developments of major economic importance occurred. The first of these was the discovery by Marais that fluoroacetic acid was the toxic constituent of the very poisonous plant known as 'gifblaar' *(Dichapetalum cymosum)*, long recognized as a serious hazard to cattle in South Africa (see p. 13). This discovery provided the incentive for biochemists and pharmacologists to investigate the mode of action of the poison and hence to attempt to develop prophylactics and antidotes; the former is now understood in its essential details, but so far no really effective cures have been evolved. The second development centred around the use of sodium fluoroacetate (Compound 1080) as a rodenticide and general mammalian pest-control agent, the preparation and uses of which are described briefly in Chapter 5 (p. 166), together with some precautions concerning its handling.

Since the war, chemical researches have continued, notably in Britain, Canada, Hungary, Israel, Japan, and Russia. However, it will surely be conceded by all that the major rôles in this period have been played by the biochemist and by the pharmacologist. Outstanding advances have been made in elucidating the mode of action of fluoroacetates, particularly by Peters and colleagues in Britain and by Chenoweth and colleagues in the United States. It is now generally accepted that fluoroacetic acid, after initial activation by formation of the thiol ester of coenzyme A, is converted *in vivo* to fluorocitric acid, which in turn blocks the tricarboxylic acid cycle by inhibiting the enzyme aconitase. It is considered likely that death results from the failure of the energy supply, with resultant impairment of cellular functions, permeability effects, etc. This aspect of the work is described more fully in a later section (p. 38).

Many of the compounds discussed in the above historical survey are listed in Table VI (p. 62).

PREPARATIONS AND REACTIONS

The most widely used procedure for obtaining the fluoroacetates involves the reaction of methyl or ethyl chloroacetate with anhydrous potassium fluoride at an elevated temperature in an autoclave[62, 78, 149]. The reaction may also be carried out at normal pressure either in molten acetamide[10] or without a solvent[112]. It is a surprising fact that sodium fluoride is almost useless as a substitute for potassium fluoride[62, 149], although antimony fluoride may be used in some instances[171].

$$ClCH_2COOCH_3 + KF \rightarrow FCH_2COOCH_3 + KCl$$

The resultant methyl or ethyl fluoroacetate can readily be hydrolyzed to sodium fluoroacetate by cold, aqueous sodium hydroxide. The industrial manufacture of sodium fluoroacetate (Compound 1080) by this means is described in Chapter 5 (p. 168); and carbon-

References p. 68

labelled sodium fluoroacetate has been obtained[154] on a micro-scale by modifications in technique. More recently, Gryszkiewicz-Trochimowski *et al.*[61] have described a convenient method of preparing fluoroacetic acid using standard, laboratory glassware: this involves the reaction of cyclohexyl bromoacetate with potassium fluoride at 200° for 48 hours; the resultant cyclohexyl fluoroacetate (74% yield) is hydrolyzed quantitatively to sodium fluoroacetate. The free fluoroacetic acid, a colorless, low-melting solid, may be obtained from the sodium salt by distillation with 100% sulphuric acid.

Fluoroacetamide may be prepared directly from chloroacetamide by halogen exchange[4, 29]. Hydrolysis of the product yields fluoroacetic acid; hence these reactions afford an alternative procedure for obtaining fluoroacetates:

$$ClCH_2CONH_2 \xrightarrow{KF} FCH_2CONH_2 \xrightarrow{H_2O} FCH_2COOH$$

Apart from halogen exchange, fluoroacetic acid or its esters may be obtained by two special methods: (a) fluorination of acetyl fluoride (using elementary fluorine diluted with nitrogen) and adding the product to ethyl alcohol[108]:

$$CH_3COF \xrightarrow{F_2} FCH_2COF \xrightarrow{C_2H_5OH} FCH_2COOC_2H_5$$

and (b) the reaction of formaldehyde, carbon monoxide and hydrogen fluoride at 160° and 750 atmospheres[1]:

$$HF + CH_2O + CO \rightarrow FCH_2COOH$$

The key reaction in preparing fluoroacetates is of course the introduction of the fluorine. Once this has been achieved, the great stability of the C-F link allows of a wide variety of common organic reactions without loss of fluorine. A few of these are shown below to illustrate the general scope:

$$FCH_2COOCH_3 + NaOH \rightarrow FCH_2COONa + CH_3OH \quad (149)$$
$$FCH_2COOCH_3 + NH_3 \rightarrow FCH_2CONH_2 + CH_3OH \quad (22)$$

$$FCH_2COOC_2H_5 + BrCH_2COOC_2H_5 \rightarrow C_2H_5OOCCHFCH_2COOC_2H_5 + HBr \quad (12)$$

$$2FCH_2COOCH_3\ (Ac_2O_2) \rightarrow CH_3OOCCHFCHFCOOCH_3 + 2H \quad (87)$$

$$C_6H_5CHO + FCH_2COOH \rightarrow C_6H_5CH{=}CFCOOH + H_2O \quad (11)$$

$$FCH_2CONH_2 - H_2O\ (P_2O_5) \rightarrow FCH_2CN \quad (22)$$

$$FCH_2COONa + BrCH_2COC_6H_4Br \rightarrow FCH_2COOCH_2COC_6H_4Br + NaBr \quad (140)$$

$$FCH_2COONa + C_6H_4\ (COCl)_2 \rightarrow FCH_2COCl + NaCl + C_6H_4\ (CO)_2O \quad (121)$$

$$FCH_2COCl + HOR \rightarrow FCH_2COOR + HCl \quad (121)$$

$$FCH_2COCl + NaOCOCH_2F \rightarrow (FCH_2CO)_2O + NaCl \quad (149)$$

$$FCH_2COCl + C_6H_6\ (AlCl_3) \rightarrow FCH_2COC_6H_5 + HCl \quad (13, 63)$$

Because sodium fluoroacetate is commercially available, it forms a very convenient starting material for preparing many different types of fluoroacetates. Detailed instructions for its conversion to fluoroacetic acid (93.5%), to ethyl fluoroacetate (90%) and to fluoroacetyl chloride (95%) are therefore given in Appendix II. The last compound is a particularly valuable fluoroacetylating agent, as can be seen from the above examples.

Of no less importance than methyl fluoroacetate is the related compound 2-fluoroethanol. It can be prepared conveniently by the autoclave technique using ethylene chlorohydrin and anhydrous potassium fluoride[152]:

$$ClCH_2CH_2OH + KF \rightarrow FCH_2CH_2OH + KCl$$

Many laboratories however do not possess an autoclave suitable for this reaction; hence the alternative procedure of Hoffmann[73] is particularly convenient. This entails the same overall reaction, but is carried out in conventional glass apparatus at 180° with vigorous stirring using ethylene glycol and diethylene glycol as solvents. The product is removed continuously by fractional distillation. The reaction may be carried out at a lower temperature if the reactants are irradiated with ultra-violet light[113].

Other methods of preparing 2-fluoroethanol have been reviewed[124]. Of these, the most interesting involves the reaction of ethylene oxide with hydrogen fluoride[90, 91, 156, 157]:

$$(CH_2)_2O + HF \rightarrow FCH_2CH_2OH$$

References p. 68

It has been reported that a small quantity of water catalyzes the reaction. While it is possible that this method could be developed on an industrial scale, it is unlikely that it or any of the other reactions could compete with Hoffmann's procedure as a convenient laboratory method of preparation.

2-Fluoroethanol undergoes most of the reactions of ethanol. The fluorine atom is unaffected by all but the most drastic treatment. Some typical reactions are shown below:

$$FCH_2CH_2OH + HOCOR \rightarrow FCH_2CH_2OCOR + H_2O \quad (121)$$
$$FCH_2CH_2OH + O \rightarrow FCH_2CHO + H_2O \quad (152)$$
$$FCH_2CH_2OH + SOCl_2 \rightarrow FCH_2CH_2Cl + SO_2 + HCl \quad (152)$$
$$FCH_2CH_2OH + SO_2Cl_2 \rightarrow (FCH_2CH_2O)_2SO_2 + 2\,HCl \quad (152)$$
$$3FCH_2CH_2OH + PBr_3 \rightarrow 3FCH_2CH_2Br + H_3PO_3 \quad (152)$$

Many other fluoroacetates and related compounds have been obtained from the corresponding chloro compounds using the autoclave technique. Primary, secondary and tertiary fluorides have been successfully prepared in this way.

The Hoffmann procedure has also been widely used. Yields are good with primary fluorides, low with secondary fluorides and negligible with tertiary fluorides[74]. The products obtained by these two procedures have been used as intermediates in the synthesis of a wide variety of compounds.

Some representative fluoroacetates are listed in Table VI (p. 62), and a few preparations and chemical reactions are summarized in Appendix I.

PROPERTIES

Physical properties

Boiling points of many of the fluoroacetates are listed in Table VI (p. 62). These afford a rough measure of volatility and hence of the danger of inhaling lethal quantities. The infra-red absorption spectrum of ethyl fluoroacetate has been recorded[104].

The figures presented in Table IV[31] summarize the physical properties of the fluorine atom in fluoroacetic acid; the corresponding data for acetic acid and the other halogenated acetic acids are included for comparison. The relative influence of the different halogens on the strength of the acids is indicated by the dissociation constants.

TABLE IV

PHYSICAL PROPERTIES OF XCH_2COOH

XCH_2COOH	*Atomic radius of X A (128)*	*Internuclear distance of X-C A (159)*	*Bond energy of X-C kcals./mole (127)*	*Electronegativity of X (127)*	*Dissociation constant of XCH_2COOH $K_a \times 10^{-5}$ (176)*
HCH_2COOH	0.29	1.14	87.3	—	1.8
FCH_2COOH	0.64	1.45	107	4.0	217
$ClCH_2COOH$	0.99	1.74	66.5	3.0	155
$BrCH_2COOH$	1.14	1.90	54	2.8	138
ICH_2COOH	1.33	2.12	45.5	2.5	75

The following random observations may be made in the light of these figures. (a) It is clear that the carbon-fluorine bond is both shorter and stronger than the other carbon-halogen bonds. Hence fluorinations involving halogen* replacements result in compounds of enhanced stability. Indeed, the outstanding stability of organic fluorine compounds has been stressed repeatedly throughout this monograph. (b) It is interesting that the toxicity of the three non-fluorine-containing halogenated acetic acids follow the order $ICH_2COOH > BrCH_2COOH > ClCH_2COOH$[111]. The toxicities are thus inversely proportional to the electronegativities of the corresponding halogen atoms; this fact has been correlated with the relative activities of the three acids as thiol acetylating agents. Extending this argument, it might be expected that fluoroacetic acid, if operating similarly, would be the least toxic of the four acids, whereas it is in fact the most toxic. Clearly then it acts by a completely different mechanism. (c) This different mechanism is associated with the atomic radii of fluorine and of hydrogen; these

* See footnote, p. 120.

are sufficiently alike as to result in a remarkable spatial similarity of the FCH_2- and CH_3- radicals. It follows that fluoroacetic acid, FCH_2COOH can mimic acetic acid CH_3COOH in its biochemical conversion to acetyl-coenzyme A and subsequent formation of citric acid; in short, fluorocitric acid is formed, which is now considered to be the ultimate toxic agent.

Chemical properties

As a consequence of the physical properties of the carbon-fluorine bond discussed above, fluoroacetates, in contrast to the other monohalogenated acids, behave very much in the same way as the corresponding unsubstituted compounds. This is a direct result of the outstanding stability of the carbon-fluorine bond[7, 149], which is unaffected by a wide range of reagents. For example, fluoroacetic acid may safely be distilled from 100% sulphuric acid at 170°[63, 149], and even after boiling with 20% potassium hydroxide for 20 hours, only 50% of the fluorine is released as fluoride[149]. It must be stressed however that aqueous solutions of various fluoroacetates (including sodium fluoroacetate, methyl fluoroacetate and sodium 4-fluorocrotonate) deteriorate markedly with time[31]. Under physiological conditions of pH and temperature, no loss of fluorine from fluoroacetate could be detected[16]. In plants, fluoroacetic acid is apparently stable for an indefinite period[25]; this is of course obvious in the special case of gifblaar (p. 13). In mammals, the carbon-fluorine bond remains intact during the formation of fluorocitric acid. One report[107] records increased storage of fluoride in the bones of rats after prolonged dosage over several weeks with sublethal quantities of sodium fluoroacetate, indicating rupture of the carbon-fluorine bond; however, the authors did not state whether their sample of 90% commercial sodium fluoroacetate was free from sodium fluoride, which if present might easily account for this observation.

There is little point in listing the large number of chemical reactions reported in the fluoroacetate series, since these may generally be predicted on the basis of the reactions of the simple unsubstituted acetates. Some general reactions of monofluoro compounds are reviewed in Appendix I. The one property apparently common to all the toxic members is the ability to form fluoroacetic acid* *in vivo*. This is developed in the next section.

Correlation between chemical structure and toxicity

Some of the key compounds prepared by Saunders and coworkers are listed on p. 20, and gross toxicity figures for many others are shown in Table VI (p. 62). As early as 1943 the main features correlating structure and toxicity had been recognized[105]. These may be summarized in the form of a simple rule: *any compound which can form fluoroacetic acid* by some simple biochemical process is toxic*. The corollary is equally significant, that compounds which cannot form fluoroacetic acid are usually non-toxic, irrespective of any superficial similarity to toxic members that they may possess. The rule and corollary may be used to advantage in predicting the toxicity of new or unknown compounds.

Some of the most important aspects of the work may be underlined at this point. Esters, salts, amides, acid halides and the anhydride of fluoroacetic acid are all toxic, on the basis of their hydrolysis *in vivo* to the parent acid. However, even minor alterations in the molecule result in inactive compounds, for example, those represented by the following formulae: $F_2CHCOOH$, F_3CCOOH, RCHFCOOR and $R_2CFCOOR$. Compounds which can form fluoroacetic acid by oxidation are as toxic as the acid itself; examples of two such compounds are 2-fluoroethanol, FCH_2CH_2OH and fluoroacetaldehyde, FCH_2CHO. Moreover, compounds which can form 2-fluoroethanol *in vivo* by some simple biochemical process are active, for example 2-fluoroethyl acetate, $FCH_2CH_2OCOCH_3$, whereas more stable derivatives of 2-fluoroethanol are inactive, for example 2-fluoroethyl chloride, FCH_2CH_2Cl. Thus the pronounced difference in pharmacological activity between 2-fluoroethanol and 2-fluoroethyl chloride is

* On the basis of more recent work, it is probably more accurate to consider the toxic species as fluoroacetyl-coenzyme A rather than free fluoroacetic acid itself.

simply due to the fact that the former can be oxidized biochemically to fluoroacetic acid whereas the latter apparently cannot[152]. ω-Fluoroacetophenone, $FCH_2COC_6H_5$ for the same reason is non-toxic[13]. The fact that 2-fluoroethyl iodide[123] is much more toxic than 2-fluoroethyl chloride may indicate that the C-I bond is more susceptible than the C-Cl bond to hydrolytic fission.

It is of interest to compare the toxicity of the two isomeric acid halides, fluoroacetyl chloride, FCH_2COCl and chloroacetyl fluoride, $ClCH_2COF$. The former on hydrolysis forms fluoroacetic acid and consequently is toxic, but the latter forms chloroacetic acid and is therefore non-toxic[149]. That the -COF grouping contributes little or nothing to the general toxicity is shown by the innocuous nature of acetyl fluoride, CH_3COF and ethyl fluoroformate, FCOOEt. Such hydrogen fluoride as is liberated during the hydrolysis is in too small amount to have any pharmacological effect.

A few compounds with unusual physiological properties may be touched upon. (a) *Fluoroacetyl salicylic acid ('Fluoroaspirin')*. *o*-$C_6H_4(COOH)OCOCH_2F$ has been reported to be as toxic as fluoroacetic acid but to cause mice to die without convulsions in drugged sleep[149]. This effect is of interest when related to the analgesic activity of aspirin; but it should be mentioned that when this compound was examined in another laboratory[33], typical fluoroacetate convulsions were found to develop. (b) *Bis-2-fluoroethyl phosphorofluoridate*, $(FCH_2CH_2O)_2POF$ was prepared[30] in an attempt to combine the toxic effects of the fluoroacetates and of the phosphorofluoridates ('fluorophosphonates'). It caused myosis as expected, and, in addition, it produced in rats a remarkable state of hyperactivity (which caused them to bite the legs of their companions), followed by atypical convulsions, coma and death. (c) That *diethyl 2-fluoroethylphosphonate*, $FCH_2CH_2PO(OEt)_2$ was non-toxic[151] leads to the suggestion that the C-P bond is not readily broken *in vivo*; similarly it may be concluded that the thioether link in '*sesqui-fluoro-H*', $F(CH_2)_2S(CH_2)_2S(CH_2)_2F$ is stable *in vivo* on the grounds that it too is non-toxic[150]. (d) *Triethyl-lead fluoroacetate*, $FCH_2COOPbEt_3$ was shown[149] to combine the sternutatory properties associated with the trialkyl-lead salts[106] with the convulsant properties of the fluoroacetates. (e) *Fluoroacetylcholine bromide*, $FCH_2COOCH_2CH_2NMe_3Br$[17, 61] has been reported[147] to have parasympathomimetic properties, characteristic of acetylcholine, and toxic properties, associated with the fluoroacetates. It thus provides a further example of combined pharmacological action. (f) Whereas simple fatty acid esters of 2-fluoroethanol are toxic, it has been found[109] that *2-fluoroethyl methanesulphonate*, $FCH_2CH_2OSO_2CH_3$ and *2-fluoroethyl p-toluenesulphonate*, $FCH_2CH_2OSO_2C_6H_4CH_3$ were devoid of toxic properties. The conclusion may therefore be drawn that these esters are peculiarly resistant to hydrolysis. However, 4-fluorobutyl methanesulphonate, $F(CH_2)_4OSO_2CH_3$ is as toxic as the parent 4-fluorobutanol, indicating that this resistance may be limited to the low members of the series. (g) One explanation for the high toxicity of the ω-fluorocarboxylates (p. 94) is that the long-chain members, because of their greater lipid solubility, penetrate the cells more readily, there giving rise ultimately to a higher concentration of fluoroacetyl-coenzyme A. As a possible means of verifying this idea, various *long-chain alkyl fluoroacetates* were prepared[121]; it was argued that, if valid, these esters should have toxicities comparable to those of the long-chain ω-fluorocarboxylates. However, no enhancement of toxicity was apparent[121] (see Table VI, p. 62), indicating that hydrolysis[15] of the esters had occurred before the increased lipid solubility had had time to exert its effect.

From all the work described above, it is abundantly clear that fluoroacetic acid is the key compound in any discussion of the toxicology, pharmacology and biochemistry of the fluoroacetates. Hence all subsequent remarks relate to it or its simple salts or esters.

TOXICOLOGY OF FLUOROACETIC ACID

It would be difficult for the author to improve upon the excellent reviews of Chenoweth[31] (covering the literature to 1949) and of Peters[133] (extending the coverage to 1956). The former is outstanding for a very complete and critical description of the pharmacology of the fluoroacetates, and the latter for a detailed discussion of the biochemistry and mode of action. The author acknowledges with gratitude permission to present in the following brief account some of the important conclusions arrived at in these two reviews, to which reference should be made for a more thorough treatment of the subject and for a complete bibliography. Other more limited reviews have appeared from time to time[68, 93].

The most outstanding feature of the toxicology of the fluoroacetates is the extraordinarily wide variation in response between different species of animals in regard both to sensitivity and to symptoms[35]. This has already been mentioned on p. 3. From an examination of the LD_{50} figures listed by Chenoweth[31], examples of which are presented in the abridged Table I (p. 3), it is apparent that the dog is sensitive to as little as 0.06 mg/kg, the rabbit to 0.5 mg/kg, and the rat to 5 mg/kg, whereas the toad is not killed by massive doses exceeding 500 mg/kg. Even in the same animal species, strain difference can account for a wide spread of values: thus, the LD_{50} for mice can vary from 0.35 mg/kg *(Microtus montebelli)* to 12.5 mg/kg (mice of mixed breed)[92]. The symptoms of poisoning are equally varied. The major point of attack may be either the central nervous system or the heart, and death may result from (a) respiratory arrest following severe convulsions, or (b) gradual cardiac failure or ventricular fibrillation, or (c) progressive depression of the central nervous system with either respiratory or cardiac failure as the terminal event[36, 39]. All these responses follow a long and essentially irreducible latent period after the administration of the poison. It is interesting to note that sodium fluoroacetate is over one hundred times more toxic than sodium fluoride (orally to rats)[172]; in man, the symptoms produced by the two compounds are surprisingly similar in many respects[6, 93], including vomiting, salivation, epileptiform convulsions, respiratory and cardiac failure, and death.

To add emphasis to these remarks, detailed symptoms are now described, following an injection of a simple fluoroacetate to a rabbit, a dog, a monkey and a rat. Chenoweth, from whose review[31] these descriptions are quoted, has concluded that these four species seem to encompass the major variations in types of response. Several workers have observed that the response of adult man to fluoroacetate may be very similar to that of the rhesus monkey; hence the description below is particularly significant to the subject of this monograph. Further reference to the effects on children and adult man is made on p. 47.

Cardiac effect: Rabbit

After an intravenous injection of sodium fluoroacetate (0.5 mg/kg) in white rabbits, no change in the animal is discernible for about one-half hour. The first effect noted is usually a weakness of the neck and front legs and a decrease in activity. This state may progress to a marked extent but usually remains moderate until the occurrence of a sudden, violent convulsion of a clonic nature, typically associated with a cry. Opisthotonus, mydriasis and blanching of the retina rapidly develop, followed by progressive relaxation, a few gasping respirations and death. If the thorax be opened immediately, the auricles are found to be beating and the ventricles usually fibrillating.

Effect on central nervous system: Dog

The onset of fluoroacetate-induced effects (usually 4 to 5 hours after 0.1 mg/kg) in the dog is heralded by a few minutes of barking and howling, 'absence' (non-recognition of human presence), actions suggestive of fearful hallucinations, hyperactivity and finally a tonic spasm followed quickly by running movements. Tonic spasms and running movements may alternate or even completely cease, and the dog may appear normal at times; but ultimately the repeated anoxic assaults on the respiratory centre during convulsions result in respiratory paralysis. The heart is often markedly slowed during convulsive seizures but rarely ceases activity until some time after the respiration has ceased. Death is typically the result of the effects of repeated and prolonged convulsions on the respiratory centre, and never primarily cardiac in origin.

Mixed response: Rhesus monkey

One or two hours after administration of the poison, the animal may vomit and becomes apprehensive and seclusive. A few minutes later, actions suggestive of auditory hallucinations are followed immediately by nystagmus. Twitching of the facial muscles, often unilateral, heralds the onset of the convulsive seizure, which is strikingly epileptiform. It quickly spreads to involve the pinnae and the masseter muscles. Spread of the convulsive activity over the rest of the body is then very rapid, ending in a jerking, symmetrical convulsion in which the spasmodic, violent jerks may occur at a rate of three per second. Tonic components are seen, but do not dominate the pattern as they do in the dog. The animal is apparently unconscious during this period; but, as the seizure passes off, it will gradually attempt to regain its feet and ultimately does so about 30 minutes after the onset of the attack. The monkey appears depressed for some time but often recovers entirely from the convulsion. A complete second seizure is infrequently seen. Generally the animal becomes weaker during the next few hours, but is often standing or otherwise exerting itself when suddenly stricken by ventricular fibrillation and death. Spontaneous recovery from ventricular fibrillation in the monkey is uncommon.

Respiratory depression: Rat

Although convulsions of a tonic nature, preceded by a one- or two-hour period of decreasing activity associated with hypersensitivity to external stimuli, are the usual result of the injection of 5 mg/kg of sodium fluoroacetate in rats of unspecified ancestry, death is the result of respiratory depression which gradually occurs long after convulsive activity has decreased or entirely ceased (see Frontispiece). Very large rats occasionally develop ventricular fibrillation. Rats which have survived an LD_{50} dose of fluoroacetate for 24 hours differ from most species (which are usually completely recovered in this time if they are to survive at all) in that they are still markedly affected. Disinclination to move is immediately apparent and is probably caused by a gross intention tremor which appears when the animal is forced to move. An extreme bradycardia can be detected by palpation or electrocardiographically. Complete recovery, if it is to occur, usually results within 48 to 72 hours after poisoning.

Among the warm-blooded species, it has been observed that primates and all types of birds are generally the least susceptible to fluoroacetate poisoning, whereas carnivora and wild rodents are most sensitive. Chenoweth[31] has drawn attention to the general tendency for herbivorous animals to manifest cardiac effects and for carnivores to develop central nervous system convulsions or depression; as might be concluded from this, in omnivorous species both the heart and central nervous system are usually affected. No appreciable sex difference in respect to sensitivity has been noted[43, 115]; however, pregnant and lactating female rats required about half again as much poison to kill as did other rats, and young rats were killed by milk from poisoned mothers[177]. Elevated environmental temperatures increase the sensitivity of mice to fluoroacetate; by the same token, frogs are more sensitive in summer than in winter[18]. These observations may be related to the lowering of body temperature[165] induced by fluoroacetate.

Although outside the scope of this monograph, the following additional comments may be made. Cold-blooded vertebrates are generally insensitive to fluoroacetate[31]. It has been reported that fish are unaffected by fluoroacetate dissolved in the water in which they swim[88]. Fleas are killed by feeding on poisoned rats[98]. Indeed, most insects, including *Anopheles* larvae[42] are generally very sensitive to fluoroacetate[67, 75, 156–158, 169] (see p. 175); this, together with the fact that fluoroacetates are readily absorbed by, but are non-toxic to plants, offers attractive possibilities for fluorine-containing systemic insecticides (p. 173). The effect of fluoroacetate on moulds, bacteria, and viruses has been studied (see p. 186), but with no outstanding results. It has been reported[58] that sodium fluoroacetate does not interfere with the guinea pig test for the diagnosis of plague (*P. pestis*).

PHARMACOLOGICAL ASPECTS

Absorption and distribution

Gryszkiewicz-Trochimowski, Sporzynski and Wnuk[63] have concluded that fluoroacetates are readily absorbed by all common routes of administration; these include inhalation, injection (subcutaneous, intraperitoneal, intramuscular, and intravenous), oral administration, percutaneous application, and introduction into the eye. More recently, some doubt has been cast upon the ready percutaneous absorption of the low members[49], particularly of solids such as sodium fluoroacetate[78], although longer chain compounds, on account of their enhanced lipid solubility, are unquestionably absorbed by this route[75]; it has been stated that of the low members, fluoroacetyl fluoride, FCH_2COF is particularly active percutaneously[63]. Even solid fluoroacetates, in the form of dusts, are effective by inhalation due to efficient absorption through the pulmonary epithelium[2, 3, 177], although most work has been carried out with volatile members. The compounds may be administered in any of the common non-toxic solvents with equal effectiveness[63]; a report[23] that methyl fluoroacetate is less toxic when injected in propylene glycol than in saline solution has not been confirmed[125].

After absorption, distribution of fluoroacetate in the rat has been found to be rather uniform throughout the animal except that the liver shows a considerably lower concentration than the other organs[65]. It has been reported[69] that, in man, the distribution of fluoroacetate was relatively uniform in the liver, brain, and kidney. Some evidence has been presented[65] that the rat can metabolize fluoroacetate to a limited extent: this conclusion is based on the findings that (a) the amount of sodium fluoroacetate recovered from the entire animal plus that excreted is distinctly less than that administered, and (b) surviving animals showed a much lower tissue concentration than did those that died. Independent evidence for the metabolism of fluoroacetate is implied by the reports that there is increased storage of fluoride in the bones of rats treated with sublethal quantities of the poison[107] (but see qualifying remarks on p. 26), and that fluoroacetate can be incorporated into non-saponifiable lipids in the rat[139].

Excretion

There is no evidence to suggest that fluoroacetate is detoxified in the body; indeed, tissues of poisoned animals retain their toxicity for long periods of time[155]. Hence urinary and faecal excretion are the only known routes for the removal of the poison[65], but reports on these are conflicting. Hagan, Ramsey and Woodard[65] have shown that rats, after receiving 5.8 mg/kg of sodium fluoroacetate, had excreted only 1% of the poison after 5 hours; with a lower dose the amount excreted after 48 hours had increased to 12%. Peters[129] recounts that Adrian took a dose of 0.65 mg/kg and produced a urine toxic to guinea pigs, the inference being that fluoroacetate (or some related compound) can be excreted by man; the same conclusion was reached by Harrisson and colleagues[69], on the basis of the fluorine content of urine obtained from a victim of fluoroacetate poisoning (see p. 51). Other workers[51] however have concluded from bio-assays with rats that sodium fluoroacetate is not excreted to any significant extent. The balance of evidence lies in favour of slow excretion after sublethal doses (with lethal doses, death prevents appreciable excretion of the poison).

Latent period

One of the most characteristic features of fluoroacetate poisoning is the invariable occurrence of a latent period prior to the onset of toxic symptoms. This is seldom of less than two hours duration, and is frequently greater. Even with large doses of poison, this is not reduced by more than a few minutes[141]. Massive doses of sodium bicarbonate, fumarate or chloride administered prior to injection of sodium fluoroacetate in rabbits decrease but do not eliminate the latent period[38]; other drugs have been examined[31, 32] which have produced dramatic shortening of the latent period in a few isolated species, but the effect is apparently not general. Longer chain members, notably sodium 4-fluorocrotonate, $FCH_2CH{=}CHCOONa$, have a distinctly shorter latent period;

this may be explained in terms of more facile cell penetration resulting from increased lipid solubility and decreased proton dissociation[31].

The latent period associated with sodium fluoroacetate is one of the many properties which contribute to the effectiveness of Compound 1080 as a rodenticide; a lethal dose may be ingested by the rat long before warning is given by the development of symptoms.

As implied above, this variable latent period may be associated with ease of cell penetration; but probably more significant in arriving at an explanation would be a kinetic study of the biochemical conversion of fluoroacetate to fluorocitric acid, the ultimate toxic agent (p. 42). In the opinion of the writer, the latent period is a result of (a) the time required for translocation and cell-penetration; (b) the time required for the biochemical synthesis of a lethal quantity of fluorocitrate; and (c) the time required for the fluorocitrate to disrupt intracellular processes (p. 43).

Tolerance

There is little doubt that repeated small doses of fluoroacetate increase resistance in rats to subsequent challenging doses[49, 85, 107], although the effect is short-lived[107]. The development of tolerance has also been reported to occur in mice[141] and possibly in the rhesus monkey[31]. On the other hand, Steyn has concluded[162] from his work with gifblaar (p. 13), of which the toxic constituent is now known to be fluoroacetate, that the continuous ingestion of the leaves or underground stems by rabbits did not induce the development of tolerance to this plant; more recently Chenoweth has reached this same conclusion for rabbits and extended it to include dogs[31]. In short, the effect varies markedly between different species. No information on tolerance is available for man.

Cumulation

Like the development of tolerance, cumulation has been observed in some species and not in others; thus, it has been observed in dogs[49], guinea pigs[49] and rabbits[162], but not in rats[49] and mice[141].

In the case of the rabbit, it has been suggested that, in addition to the cumulative effect, repeated small doses of fluoroacetate may result in enhanced sensitization[162]. As an example of the cumulative effect in dogs, Foss[49] has reported that the animals succumb in 4 or 5 days to daily administration of one-quarter of the lethal dose, whereas they survive indefinitely if the same or even a larger dose be given on alternate days. From this one may conclude that the poison is being eliminated very slowly, as was suggested in the discussion of excretion (p. 35); further, Chenoweth[31] has used these data to infer that at least some enzyme-fluoroacetate combinations are reversible, as is apparent also from the fact that animals which have received sublethal doses ultimately recover completely. No information on cumulation is available for man.

Antidotes, antagonists and prophylactics

No highly effective antidote to fluoroacetate poisoning has yet been found. From a consideration of the mechanism of action (p. 42), it is apparent that any C_2 precursor might afford protection by selective competition with fluoroacetate for the active sites; however, solubility and penetration effects have reduced drastically the number of successful candidate agents. For example, sodium acetate itself[170] and ethanol [77] have both been found to be effective, but only in certain animals; it is of interest that, at least in mice, the two compounds combined are distinctly more active than either alone[77, 170], suggesting some synergistic effect. In one case of fluoroacetate poisoning in man (Non-fatal case 1, p. 48), hourly dosage with 100 proof whisky mixed with sugar and water apparently provided some relief[178]. No antidotes have been discovered[133] that will remove or inactivate the fluorocitrate (p. 42) once it has been formed. Barbiturates may be used to control convulsions, but Foss[49] has warned that in some animals such treatment may only accelerate death by increasing the respiratory depression; the application of intravenous anaesthetics has to be pushed to full doses to produce anticonvulsant effects. In short, it may be stated that of the very many compounds examined[31, 133], only two (monoacetin (glycerol monoacetate, glyceryl monoacetate)[37] and acetamide[55, 56]) are outstanding but by no means universally effective. In monkeys, which are apparently similar to man in their response to fluoroacetate, monoacetin therapy significantly reduced mortality and alleviated the symptoms of intoxication[37], even so long as 30 minutes after injection of the poison, when convulsions had already started to develop. Acetamide has been examined only in guinea pigs and rats; it was found to be effective if administered up to 8 minutes after injection of fluoroacetate. In the opinion of the author, the balance of evidence slightly favours monoacetin as the preferred antidote and prophylactic.

Medical aspects are discussed in a later section (p. 47). The recommended treatment involves immediate emesis, the *judicious* administration of short-acting barbiturates to control convulsions, and monoacetin therapy. Instructions are given in Appendix III (p. 208) for easy reference.

BIOCHEMICAL ASPECTS: MODE OF ACTION

The reader is referred to the excellent reviews of Chenoweth[31] and of Peters[129, 131, 133] for an exhaustive treatment of this subject. The following account provides a summary of the historical developments which led up to the mechanism of fluoroacetate poisoning as it is understood today. As in any field in which frequent new discoveries are being made, it is possible that some of the conclusions will require modification in the light of future results.

The story started in the early years of the war when McCombie and Saunders observed that no enzyme system had been found which was inhibited *in vitro* to any extent by methyl fluoroacetate[105]; this has since been confirmed by other workers[81, 129, 133]. No fresh evidence was forthcoming until 1947 when Bartlett and Barron[7] suggested, from experiments on the metabolism of animal tissues, that fluoroacetate exerted its toxic action by inhibiting the conversion of acetate to 'active acetate' (now recognized as acetylcoenzyme A) prior to its oxidation in the tricarboxylic acid cycle; although this explanation is now considered to be incorrect, the preoccupation with a mechanism involving C_2 fragments has been vindicated by later work. Shortly afterwards, Kalnitsky and Barron[83] observed that citric acid accumulated markedly during experiments involving the action of fluoroacetate on rabbit kidney homogenates. In the same year, Liébecq and Peters[94] independently recorded this very important symptom of fluoroacetate poisoning, using a washed kidney mitochondrial preparation; the effect was accompanied by a decrease in oxygen uptake and no change in acetic acid concentration.

The 'active acetate' theory accounted for the competitive inhibition of acetate oxidation by fluoroacetate observed by Bartlett and Barron[7], and support for the theory, as an adjunct to the fluorocitrate mechanism (p. 42), has been voiced as recently as 1957[175]; clearly, if the acetate could not be incorporated into the tricarboxylic acid cycle, its normal metabolic route would be blocked. In an examination of this postulate, Liébecq and Peters[95] found that it was possible with fluoroacetate to block oxidation of fumarate in a guinea pig's kidney homogenate *without* accumulation of acetate; this finding invalidated the 'active acetate' theory. Moreover, in the course of this work, Liébecq and Peters, in studying the above-mentioned accumulation of citric acid, determined that it was due to a blockage of citric acid oxidation.

Up to this point, all biochemical experiments had been carried out *in vitro*; but in 1949, Buffa and Peters[26, 27] made the important discovery that citric acid accumulation occurs also *in vivo*. In their report[27], these authors recorded the widespread occurrence of this effect in the rat (Table V) one hour after receiving an intraperitoneal injection of sodium fluoroacetate (5 mg/kg). The accumulation of citric acid has since been found by many workers to occur both *in vitro* and *in vivo* in a variety of animals and animal tissues. It is important to note that, following a sublethal dose, the citric acid concentration reaches a maximum in the rat after 4 to 6 hours and becomes normal again after about 40 hours[96]; and that following a lethal dose, citric acid accumulation is negligible within about one hour after death[8].

Summarizing these findings, it was apparent: (a) that citric acid

References p. 68

TABLE V

ACCUMULATION OF CITRIC ACID IN THE RAT

	Citric acid, μg/g wet tissue Control	Poisoned
Kidney	14	1036
Heart	25	677
Spleen	0	413
Stomach	37	386
Small intestine	36	368
Large intestine	21	248
Lung	9	257
Brain	21	166
Blood	3	50*
Liver	0.8	31**
Diaphragm	0	400
Uterus (virgin)	217	207

* The elevation of plasma citrate is augmented by parathyroidectomy or nephrectomy[50].
** Fluoro*citrate* injections cause an immediate rise in the citrate content of the liver[53].

accumulated due to blockage of citric acid oxidation in the tricarboxylic acid cycle; but (b) that fluoroacetate had no action on the individual enzymes of the tricarboxylic acid cycle. This apparent anomaly was resolved by the ingenious and far-reaching suggestion of Liébecq and Peters[95] that fluoroacetate was converted to fluorocitric acid by the biochemical mechanism normally responsible for the conversion of acetate to citric acid, and that the fluorocitric acid thus formed was the ultimate toxic agent. This same idea was advanced independently at about the same time by Martius[102], and later by Elliott and Kalnitsky[45]. Such a conversion by enzymic synthesis of a non-toxic substance, fluoroacetic acid, to a toxic one, fluorocitric acid, has been termed by Peters a 'lethal synthesis'. For a proper appreciation of this mechanism, an account must now be given of the biochemistry of acetate oxidation by the tricarboxylic acid cycle.

Acetate is considered to be metabolized by the following steps:

(a) *Activation.* The long-recognized 'active acetate' is now known to be acetyl-coenzyme A (I) the formation of which occurs when free acetate and coenzyme A react in the presence of adenosine triphosphate. (b) *Incorporation into the tricarboxylic acid cycle.* The acetyl-coenzyme A (I) reacts with the enol form of oxaloacetic acid (II) to produce citric acid (III), the first member of the cycle. The reaction is catalyzed by the 'condensing enzyme'. (c) *Oxidation in the tricarboxylic acid cycle.* By the chemical changes summarized schematically in Fig. 5, the citric acid so formed is degraded through

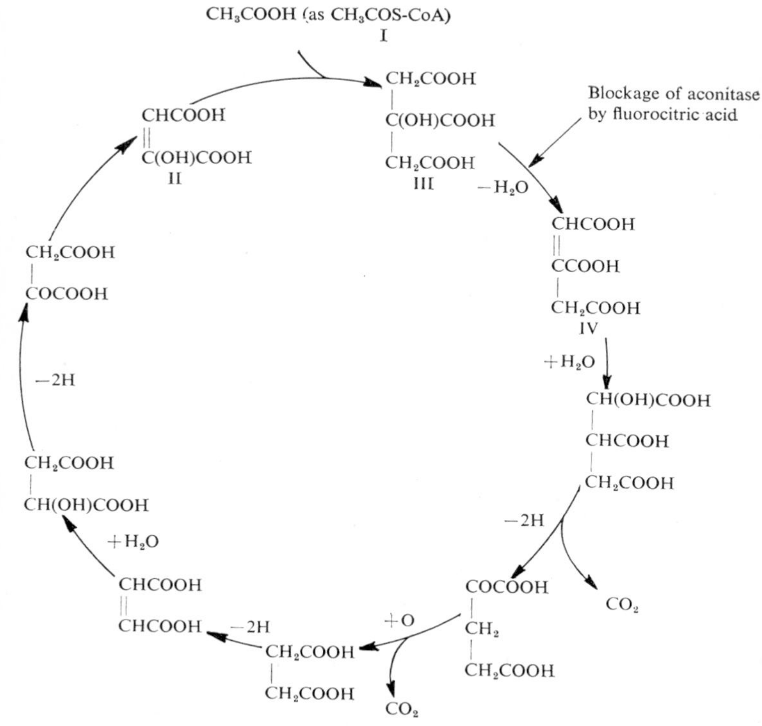

Fig. 5. The tricarboxylic acid cycle. Specific enzymes are not included. Although most of the reactions are reversible, the cycle is probably unidirectional in operation, and is shown as such in this simplified schematic representation.

References p. 68

the agency of specific enzymes to carbon dioxide, water and oxalo-acetic acid; the latter, in being regenerated, is theoretically available for the oxidation of an unlimited quantity of acetate to carbon dioxide and water. For the discussion that follows, it must be mentioned that the enzyme aconitase is responsible for the conversion of citric acid (III) to *cis*-aconitic acid (IV).

It is now certain that the overall scheme as shown below operates as an essential biological path in the normal metabolism of intact animals, the corollary of this being that interference with the process must result in gross organic dysfunction leading ultimately to death. This is of fundamental importance in the discussion of the biochemical lesion caused by fluoroacetate poisoning, and is referred to again (p. 43).

The remarkable spatial similarity of the CH_3- and FCH_2-radicals has already been stressed (p. 26). It is possible to conclude from this that acetic acid and fluoroacetic acid bear a pronounced structural resemblance one to the other. Consequently fluoro-acetic acid can mimic acetic acid in undergoing activation and incorporation into the tricarboxylic acid cycle; in short, *fluoro-citric acid* is formed.

$$FCH_2COOH \rightarrow FCH_2COS\text{-}CoA$$

$$\begin{array}{lcl} CHCOOH & & CH_2COOH \\ \| & & | \\ C(OH)COOH & \rightarrow & C(OH)COOH \\ \quad + & & | \\ FCH_2COS\text{-}CoA & & CHFCOOH \end{array}$$

But here the mimicry ceases. Instead of being metabolized further by the cycle, the fluorocitric acid acts as a potent competitive inhibitor of the enzyme (aconitase) responsible for the next step[97, 110, 132]. The inhibitory action may possibly be due to an irreversible combination with aconitase. (It can be visualized that the fluorocitric acid resembles citric acid sufficiently to enter the enzyme matrix but that the physico-chemical properties of the fluorine atom hinder its subsequent dislodgement.) Whatever the intimate mechanism, the inhibition of aconitase results in the blockage of the tricarboxylic acid cycle. As a result of this, it is likely that the energy supplied by the cycle is gradually reduced to the point where permeability barriers are destroyed, cellular functions are impaired, and death results (see p. 42). The large build-up of citric acid which accrues from this enzymic blockage may be considered as a symptom of poisoning and not the cause of death; it can be used to advantage as an indicator of the tricarboxylic acid cycle (Chapter 5). The wide variation in response to fluoroacetate in different species may be associated with the lack of some component of the activating system necessary for forming fluorocitrate[133]. It is probable that fluorocitrate, once formed, is slowly eliminated, because animals ultimately recover completely from sublethal doses, and citrate accumulations diminish with time.

It is appropriate at this point to survey some of the evidence in support of the above explanation. (a) Prior to its conversion to fluorocitric acid, fluoroacetate has been assumed to form fluoroacetyl-coenzyme A as activation for the reaction. Proof of this has been provided by Brady[19], and by Marcus and Elliott[101], who have synthesized and studied fluoroacetyl-coenzyme A. With oxaloacetate in the presence of crystalline condensing enzyme, it reacted to form fluorocitrate, which was estimated enzymically. It is likely that activation of fluoroacetate by formation of fluoroacetyl-coenzyme A takes place *in vivo* at some centre other than the usual one for acetate activation[131]. As might be expected, there is some evidence that fluoroacetyl-coenzyme A can undergo biochemical syntheses normally associated with acetyl-coenzyme A; for example, fluoroacetate can be incorporated into a non-saponifiable lipid[139]; moreover, Peters [133] has suggested that the long-chain fluoroacid isolated from ratsbane (p. 83) arises from fluoroacetate by a biochemical synthesis. (b) Of prime importance was the isolation and characterization of fluorocitric acid. Working with kidney homogenates poisoned with fluoroacetate, Peters and colleagues[28, 135] succeeded in isolating a tricarboxylic acid distinct from citric acid; this was shown to be extremely active in inhibiting the disappearance of citric acid from a test system composed of kidney homogenates. The isolated material was thought to be fluorocitric acid on the basis of its similarity to citric acid in paper chromatography and of the presence of fluorine in the molecule. Final proof of structure was obtained by comparing[136] the infra-red spectrum of the isolated material with that of fluorocitric acid prepared synthetically by Rivett[145]. (c) To prove that fluorocitrate is formed by a 'lethal synthesis' (p. 40), it is necessary to show that fluorocitrate is toxic under conditions where fluoroacetate is non-toxic. This has been achieved by Peters[130, 134] by injecting into the subarachnoid space of pigeons small quantities (50–80 μg) of fluorocitrate. After 10 to 20 minutes a period of agitation became apparent, terminating in violent convulsions and death. Similar injections of fluoroacetate were without effect, even when given in amounts larger than would be the equivalent of the fluorocitrate. This crucial experiment indicates that pigeon brain tissue does not form fluorocitrate from fluoroacetate; moreover, it provides a basis for inferring that, in animals poisoned in the normal way, fluoroacetate is converted (probably in the mitochondria) to fluorocitrate, which in turn is transported to the brain to produce the characteristic convulsions (but see next paragraph). Similar observations have been made with the rabbit[70] and the rat[53]. (d) Evidence that fluorocitrate is a specific competitive inhibitor of the enzyme aconitase has been provided by Judah and Rees[79] and by Peters and colleagues[110, 137, 138]. Working with a soluble, purified aconitase, it was shown that, at a concentration of 1.6×10^{-4} *M*, fluorocitrate resulted in 58% inhibition. This inhibition is considerably less than that shown with mitochondrial preparations of kidney, a fact which is not yet completely understood. So far, aconitase is the only enzyme known to be inhibited by fluorocitrate. (e) That convulsions and death can indeed result from a blockage of the tricarboxylic acid cycle has been shown by Peters[130] in work on thiamine. A deficiency of this vitamin results in a blockage of the entry of C_2 fragments into the tricarboxylic acid cycle by interference with the enzyme 'pyruvic oxidase'. Consequently this results in deprivation of the energy (derived from the tricarboxylic acid cycle) necessary to maintain the tissues in a living state. Since fluoroacetate results in a similar deprivation, the conclusion may be reached that the biochemical lesion produced by the two effects should be the same; such is indeed the case, as is evidenced by convulsions and death in both instances.

The above account is perhaps an oversimplification but it contains most of the salient features. A few points must now be raised which present difficulties as yet unresolved. (a) If fluorocitrate is indeed the ultimate toxic agent, it should be more toxic (on a molar basis) than fluoroacetate; however, such is not the case[53]. This inconsistency may well be due to the inability of fluorocitrate to penetrate vital tissues such as heart and brain[44, 133]. (b) Peters[133] has drawn attention to this very problem of the transport of fluorocitrate to the brains of intact rats. Since rat brain tissue itself apparently does not carry out the conversion of fluoroacetate to fluorocitrate, the latter must be synthesized elsewhere and subsequently enter the brain of the rat through the 'brain barrier'. Yet the low toxicity of fluorocitrate makes such penetration unlikely, as explained in (a). How then does the fluorocitrate gain entry to the brain? In suggesting an answer to this question, Peters[133] has pointed out that the apparent anomaly would disappear if cells in the 'brain barrier' were the active agents for the synthesis of fluorocitrate from fluoroacetate. It is possible however that this problem is not of general importance, since Chenoweth[34, 71] has inferred that fluoroacetate may be converted to fluorocitrate in the brain of various animals, including the rabbit[39] and the dog. (c) Lingering doubts about the fluorocitrate theory are revived by certain differences between enzymically-synthesized and chemically-synthesized fluorocitric acid. While the infra-red spectra of both are substantially the same, certain slight discrepancies are nevertheless apparent[136]. Moreover, the two materials behave differently in their inhibition of aconitase preparations (mitochondrial and purified)[110]. Tentative explanations for these inconsistencies have recently been advanced[80, 160], based on stereochemical differences of the four possible isomers of fluorocitric acid. It is appropriate to add that the synthetic material has been shown recently to contain varying mixtures of up to twelve substances[173, 174].

It is worth recording that the accumulation of citric acid was for several years considered to be the cause of tissue disturbance

References p. 68

and death, possibly due to immobilization of calcium ions[27, 129]; but Chenoweth and colleagues[71, 84, 85] have effectively disposed of this idea, at least as the primary cause. It should be mentioned too that Chenoweth, particularly in the early years[31, 84], did not consider inhibition of the tricarboxylic acid cycle to be responsible for the pathological conditions. Even today, other explanations for fluoroacetate activity cannot be entirely excluded. For example, Benitez, Pscheidt and Stone[9] have shown that a significant increase in brain ammonia occurs in animals poisoned with fluoroacetate, and it is recognized that ammonia can cause many of the typical convulsive activities which have been described for fluoroacetate; it is possible of course that this build-up of ammonia may be attributable indirectly to the action of fluorocitrate on aconitase, since blockage of the tricarboxylic acid cycle could interfere with transaminase processes at the α-ketoglutarate stage[20].

The toxic long-chain ω-fluorocarboxylic acids apparently behave in basically the same way as fluoroacetate after initial degradation to fluoroacetyl-coenzyme A (p. 91). However, certain qualitative and quantitative differences have been reported. For example, 4-fluorobutyric acid is more toxic and acts more rapidly than fluoroacetic acid, and the resultant distribution in the tissues of accumulated citric acid is different with the two compounds. These points are discussed more fully in Chapter 3.

The preparation and properties of all eight α,α-difluoro-substituted acids of the tricarboxylic acid cycle have recently been described[142]. These consisted of α,α-difluoro-oxalacetic acid, $HOOCCOCF_2COOH$; α,α-difluorocitric acid, $HOOCCH_2C(OH)(COOH)CF_2COOH$; α,α-difluoroaconitic acid, $HOOCCH{=}C(COOH)CF_2COOH$; α,α-difluoroisocitric acid, $HOOCCH(OH)CH(COOH)CF_2COOH$; α,α-difluoro-β-oxalosuccinic acid, $HOOCCOCH(COOH)CF_2COOH$; α,α-difluoro-γ-oxoglutaric acid, $HOOCCOCH_2CF_2COOH$; α,α-difluorosuccinic acid, $HOOCCH_2CF_2COOH$; and α,α-difluoromalic acid, $HOOCCH(OH)CF_2COOH$. These members had relatively low mammalian toxicity (LD_{50} 50–150 mg/kg by intraperitoneal injection into mice), but showed some fungicidal, insecticidal and miticidal activity.

MEDICAL ASPECTS

Diagnosis and case histories

From a study of the case histories that follow, it will be apparent that the major toxic effects of the fluoroacetates in man involve the central nervous system and the heart. Epileptiform convulsions alternate with coma and depression (which may involve the medulla). Cardiac irregularities and sudden cardiac arrest are prominent features. These symptoms are usually preceded by an initial latent period of up to six hours characterized by nausea, vomiting, excessive salivation, numbness, tingling sensations, epigastric pain, and mental apprehension; other signs and symptoms which may develop subsequently include muscular twitching, low blood pressure, and blurred vision.

The convulsions are often very severe. Depression may be deep and associated with failure of respiration. Ventricular fibrillation and cardiac arrest may occur suddenly. Death can result from cardiac arrest, asphyxia during a convulsion or respiratory failure. Secondary infection (particularly of the lungs) or shock may contribute to the fatal outcome. It is however the effect on the heart that is considered to be the primary cause of death; hence the cardiac status should be followed carefully, preferably by the electrocardiograph. (Evidence of arrhythmia and marked change in the shape of the T-wave are of great prognostic significance.) Chenoweth[31] has concluded that the response of man may be similar to that of the monkey (p. 32); the hospital case histories that follow support this view.

It is impossible to be specific about the toxic dose for man; a likely range lies between 2 and 10 mg/kg. That man is relatively insensitive to fluoroacetate has been shown by Adrian and by Gryszkiewicz-Trochimowski and co-workers in experiments on themselves (see pp. 4, 17).

To date there have been some 30 cases of poisoning with fluoroacetate, of which at least 16 have been fatal; some of these were successful acts of suicide, but most resulted from accidental ingestion of the poison ('drank solution of Compound 1080 from unlabelled whisky bottle', 'chewed on rodenticide bait cup', etc.). Some representative case histories follow, which have been selected to include three non-fatal and three fatal cases; all six patients were male, but of varying ages (from 13 months to 40 years). The accounts give a summary of the main symptoms and course of poisoning; methods of treatment have been omitted as being mainly symptomatic (and sometimes valueless). Reports of pathological and microscopic examinations have also been omitted, because they contain no characteristic or specific signs indicative of fluoroacetate poisoning.

Non-fatal case 1[178]. A thirteen-month-old boy (W.R.) was found with a small paper baitcup containing Compound 1080 (sodium fluoroacetate); the poison was smeared over his lips and hands. His mother washed his face and mouth with running water and brought him to hospital where he was treated by gastric lavage approximately 15 minutes after ingesting the poison. For the next 5 hours he slept soundly with good respiration and a pulse rate ranging between 80 and 96. Thereafter he developed slight muscular twitching of the fingers and toes, accompanied by a rising and irregular pulse rate. The twitchings gradually extended up his arms and the pulse rate rose to 150. Shortly thereafter he had a convulsive seizure in which he stiffened and his body jerked four or five times. For the next few minutes he was very restless, but he gradually relaxed and his pulse rate dropped to within normal bounds. The pulse rate then rose again very rapidly and twitching of his hands rapidly developed into two generalized convulsive seizures. The pulse rate at this point was too rapid to count, but was thought to be about 200. The twitchings recurred frequently for the next few minutes and the heart-beat was very fast. The condition then subsided for about half an hour, when suddenly he screamed and doubled up as though he were having severe abdominal pain. From this point onwards, he gradually recovered and slept soundly except when occasionally he showed toxic manifestations of the barbiturates which had been administered (tossing in his sleep, screams etc.).

Non-fatal case 2[52]. A previously healthy, well developed and well nourished two-year-old Negro boy (N.D.) was admitted to hospital, apparently moribund, about 6 hours after he had licked the screw top of a bottle containing a solution of sodium fluoroacetate. (The bottle was labelled in a handwritten script 'Rat Poison – 1080' and had been purchased by the parents from a passing pedlar two years previously.) Almost immediately after licking the screw top, the boy had vomited, and he continued thereafter to vomit and retch frequently. Three hours later a local physician was consulted who prescribed a cathartic of unknown nature which the child vomited. About 6 hours after ingestion of the poison he began to have generalized convulsions and became stuporous, whereupon he was brought to hospital. At this point he was in a critical condition (comatose, and exhibiting carpopedal spasm, tetanic convulsive movements, irregular respirations and great cardiac irregularity). A few hours after his admission the child became responsive and appeared to recognize his father, but he continued to vomit most of the milk given by gavage. Four hours after his admission a generalized tonic-clonic convulsion lasting several minutes occurred, which was followed by deep post-convulsive coma from which he could not be aroused. One hour later he became slightly responsive again but could not speak or sit up. Later further convulsions occurred followed by a long period of coma.

On the day following his admission, he was limp and unresponsive, and his eyes wandered about. This condition was interspersed with frequent tonic convulsions and groaning; during the convulsive spasms the pupils dilated and remained inactive to light, but at all other times they were round, regular, equal, and reactive to light. He did not respond to painful stimuli until they were of severe intensity and then he only groaned and showed no withdrawal response. Cardiac rhythm remained irregular, with exaggerated sinus arrhythmia and many dropped beats, occasional periods of asystole or of complete arrhythmia (fibrillation) and infrequent runs of what appeared to be paroxysmal tachycardia.

The cardiac rhythm and the quality of cardiac sounds changed

References p. 68

frequently during the first three days in hospital, and tonic convulsions lasting several minutes occurred many times each hour. During the third day two long periods of apnoea developed, which were relieved by artificial respiration in conjunction with an oxygen mask. Gradually however his condition improved until in the evening of the third day he was observed to whimper on painful stimulus, and in the evening of the fourth day (100 hours after ingestion of the poison) he began to open his eyes and look about. During the following hours a remarkable change in consciousness and responsiveness occurred, and he steadily regained all his motor ability, became articulate and lost his drowsiness. On the evening of the sixth day he was clinically well.

One year after his discharge, he appeared to be completely healthy; mental and physical development had proceeded normally, and there had been no further convulsions or other symptoms of neurological disorder.

Non-fatal case 3[177]. The patient, a U.S. Naval Sanitation Officer (A.T.W.) was conducting field tests with sodium fluoroacetate in the Pacific area, when a sudden gust of wind blew an undetermined amount of the powder into his face; some of it was inhaled. A tart, sourish taste was noted, followed almost immediately by a tingling sensation around the corners of the mouth and in the nasal passages. Becoming alarmed, he sought medical assistance. Soon the entire face had become numb; this condition was accompanied by an excessive flow of saliva and loss of speech. The tingling sensation rapidly spread throughout the arms and legs. Vision was blurred from the outset, with inability to focus on objects. Spasmodic contractions of the voluntary muscles occurred, followed by violent convulsions and coma (two and a half hours after ingestion of the poison). Further convulsions occurred at sporadic intervals with varied intensity for several hours. 'When first seen, the patient was in typical grand mal type epileptiform convulsion with dilated pupils, foaming and frothing at the mouth, rolling of the eyeballs, and muttering prior to and after seizures'. Carpal spasm was marked, with thumbs inverted and fingers flexed into a cone shape. The exposed skin was markedly cyanotic, and there were beads of moisture on the face, lips and forehead. Other symptoms included generalized jerking of the legs and arms; clenching of the teeth; stertorous laboured breathing with foam and thick mucus in the nose, mouth and throat; nystagmus; and irregular heart beat.

The patient gradually recovered after some rather heroic treatment. The extreme difficulty in respiration while in the prone position was eased by using a back rest at 45° and oxygen therapy. Following recovery, no after-effects became manifest.

It is of considerable interest to note that throughout the entire duration of the illness, no actual pain or discomfort was experienced, other than alarm and anxiety. The patient recalled pronounced cutaneous hyperaesthesia, in which the bed linen felt like canvas.

Fatal case 1[69]. A white man (F.K.), aged about 40 years, was found unconscious in his bedroom. According to his wife, he had never had any serious illness or operation until after discharge from the Army following World War II. At this time he was confined to hospital with battle fatigue for 6 months, and received shock treatment for severe depression. He was readmitted to hospital some 9 months later for a further 2 month period, with a diagnosis of manic depressive psychosis. On his release from hospital, his family was warned of danger of possible suicide.

He was admitted to the emergency ward with the information from his wife that he had probably taken sodium fluoroacetate. He was unconscious and had nystagmus of both eyes. Irregular heart rhythm and slight muscular spasms were noted. Shortly afterwards the nystagmus became pronounced and he had an epileptiform convulsion. Muscle spasms of the extremities continued, perspiration became profuse, and his face was flushed. He retched frequently and occasionally opened his eyes. He gradually became very restless, frothing at the mouth and experiencing difficulty in breathing because of excessive mucus secretion. At times he would moan loudly and thrash about in bed. Respiration became increasingly

References p. 68

laboured, and, some 17 hours after admission, he was pulseless, ceased breathing, and was pronounced dead.

The fluorine content of the organs and body fluid was determined, and the results obtained were used to calculate the theoretical fluoroacetate content. The distribution was found to be relatively uniform in the liver, brain and kidney. A high level in the urine indicated rapid urinary excretion. In spite of the gastric lavage which was carried out on admission to hospital and the 17 hour delay before death occurred, important quantities of the poison were found in the stomach, indicating the advisability of early and repeated gastric lavages. A total of about 465 mg of sodium fluoroacetate was calculated to be present in the organs and body fluids examined.

Fatal case 2[116]. The patient was a well developed and well nourished 15-year-old white male, who inadvertently drank an unknown quantity of Compound 1080 (sodium fluoroacetate) in water; he had evidently thought the solution to be a soft drink, because it was contained in an unlabelled 'pop' bottle. He complained of no ill effects until about 2 hours later when he left the table during his evening meal and vomited. During the next half hour he had a convulsion and vomited again. When he was first examined, he was comatose and never revived sufficiently to give any information. Some 4 hours after ingestion of the poison, he was having almost continuous epileptiform convulsive seizures and was moderately cyanotic in spite of oxygen administered by nasal catheter. His skin was moist and occasional tonic-clonic movements of isolated muscle groups were apparent. His pupils were equal and reacted to light. The only response to painful stimuli was a twitching of the muscles or slight convulsion. His pulse was rapid but regular, and his colour and general appearance were good. However the patient's condition became steadily worse, until, some 7 hours after ingestion of the poison, the apex beat was 144 while the pulse was around 72*. Half an hour later, his breathing became

* Author's note: this is an excellent example of pulsus alternans, only alternate beats being sufficiently strong to be felt at the wrist (see p. 56).

suddenly laboured and irregular; his colour turned cyanotic; his pulse disappeared entirely; and he expired shortly afterwards, apparently from ventricular fibrillation.

Fatal case 3[21]. This case is of particular interest, because the patient lived for 5 days after ingesting the poison, and extensive laboratory work was therefore possible. A 17-year-old boy, son of a professional rat exterminator, entered the emergency ward of the hospital at 4 a.m. and told the nurse that he had swallowed a solution of about 4 oz. of sodium fluoroacetate in water, after which he had promptly vomited. He stated that he had noted almost immediate epigastric pain. At the time of admission, about 1 hour after ingesting the poison, he was alert and responsive, but complained of epigastric pain. During gastric lavage, he gradually became more and more unresponsive, until he was comatose. Some 3 hours after drinking the solution he had a grand mal convulsion associated with faecal incontinence. He was now in deep coma, and unresponsive to painful stimuli. There was dusky cyanosis of the nailbeds and lips. The pupils were constricted but reacted normally to light. There was frequent chewing movements of the jaws. An electrocardiogram obtained at this time was interpreted as showing right axis deviation, ventricular premature contractions, and evidence of diffuse myocardial abnormality.

Eight hours after admission, the patient vomited some dark brown material, which gave a chemical reaction for blood. Examination at this time revealed that the cyanosis had disappeared. Coma persisted. During the next 12 hours, the patient became very restless, thrashing about in bed. There were frequent episodes of severe carpopedal spasm, while at other times all the muscles of the body became very spastic.

On the morning of the second day, acute pulmonary oedema supervened; this cleared readily on treatment, but coma persisted. The pupils were small and did not react to light, respirations were 40 per minute, and the pulse was feeble at a rate of 160 per minute. During the ensuing 4 hours, the pulse rate rose to 180 per minute and the blood pressure dropped to 85/0 mm Hg. Suction of the

References p. 68

upper respiratory tract had to be carried out frequently, and an endotracheal tube was inserted. Later in the morning, the pupils once again responded to light.

On the third day the clinical picture remained unchanged except for a further drop in blood pressure to the point of being unobtainable. The temperature continued to rise and reached a maximum of 104.6° F late in the day.

On the fourth hospital day there was a tremendous increase in the amount of tracheobronchial secretions, which necessitated almost constant suctioning. The patient's condition seemed to improve a little under treatment, and the blood pressure was obtainable at 100/60 mm Hg., although the pulse remained rapid (180 per minute) and feeble.

On the night of the fifth hospital day, the tracheobronchial secretions became so copious and tenacious that an adequate airway could not be maintained without a tracheotomy. Following this procedure, thick yellowish-white mucoid material was suctioned from the lower trachea, and the airway immediately sounded clear and dry. However, the patient's temperature was rising steadily until, by 3 a.m. on the sixth hospital day, it had reached 108° F in spite of all measures to reduce it. From this point, the respirations became extremely laboured and rapid, the blood pressure once again was unobtainable, and the pulse increased to such a rapid rate that it was impossible to count it with any degree of accuracy. At 8 a.m. the patient died.

Therapy

The reader is referred to p. 37 for a general discussion of antidotes, antagonists and prophylactics which have been investigated in connection with fluoroacetate poisoning in experimental animals.

The following instructions have very kindly been supplied by Maynard B. Chenoweth, M.D.* They are repeated in Appendix III (p. 208) for easy reference. Lacking specific information to the

* Director of Pharmacological Research, Biochemical Research Laboratory, The Dow Chemical Company, Midland, Mich.

contrary, the recommendations may be considered as general for fluoroacetates and for all compounds which owe their toxicity to the ultimate formation of fluoroacetate. The instructions may be augmented by symptomatic treatment, at the discretion of the attending physician. Common-sense procedures, such as removal of contaminated clothing and washing of the skin, have not been included.

(a) FIRST AID (N.B. *Immediate treatment is of the utmost urgency*).

1. If the poison was swallowed and the patient is conscious and not convulsing, induce vomiting immediately by either of the following methods:

(a) Drinking 2 to 4 oz. of a *strong* solution of table salt in water (*i.e.* a solution containing as much table salt (sodium chloride) as will readily dissolve).

(b) Stimulation of the back of the throat with a spoon or padded stick (such as a wooden pencil with an erasing rubber attached). A finger may be used, but only as a last resort, since there is danger of it being bitten.

2. CALL A PHYSICIAN!

3. When available, monoacetin (glycerol monoacetate, glyceryl monoacetate) of any purity (Technical or better) (100 c.c. in 500 c.c. of water, *i.e.* approximately 3 oz. in a pint of water) may be drunk. Although the taste is unpleasant and further vomiting may occur, such treatment can do no harm but may do considerable good.

N.B. NEVER force fluids by mouth to unconscious or convulsing persons!

If a long delay is unavoidable in the arrival of the physician, a second identical dose of monoacetin may be taken after about one hour.

4. Keep the patient warm and quiet.

(b) DEFINITIVE TREATMENT. *To the physician: Hospitalization should be prompt; immediate treatment is of the utmost urgency.*

On the basis of a few human poisonings, the following symptoms and signs may be expected:

References p. 68

1. *Epileptiform convulsions* at any time from 30 minutes to 48 hours after ingestion of the poison.

They may be controlled by intravenous (or intramuscular) barbiturates by the usual procedure for severe convulsive states. They are *not* the usual cause of death.

2. *Cardiac irregularities*. Variation in the rate and rhythm may become very great. Usually there is alternation, progressing to failure of every other beat to be detectable at the radial pulse (pulsus alternans). Ventricular extrasystoles occur with increasing frequency as poisoning progresses, and death is usually produced by ventricular fibrillation or cardiac arrest.

Treatment

On the basis of animal (monkey) experiments[37], it is to be anticipated that the following treatment may be beneficial. Monoacetin (glycerol monoacetate, glyceryl monoacetate) in large doses by intramuscular injection is a specific antagonist to fluoroacetate. The doses which might be employed are 0.1 to 0.5 c.c. per kg (2.2 lb.) body weight, *i.e.* 6 to 30 c.c. for an 11 stone (150 lb.) man. Some pain and oedema may be expected from such treatment. However, the toxicity of monoacetin is very low, and such slight respiratory stimulation, vasodilation, and sedation as may occur need not cause alarm.

The dosage should be repeated after about 30 minutes and whenever the patient's condition suggests progression of the poisoning. Continuous observation *by the physician* may be required for 48 hours, because deterioration of the patient may occur rapidly and unexpectedly.

Notes:

(1) If monoacetin is unavailable, acetamide dissolved in physiological saline may be used[55] by the same procedure in approximately the same dosage; however, the historical background for the use of this material is less extensive.

(2) If the poison was swallowed, immediate and thorough gastric lavage should be of great value.

(3) Oxygen and artificial respiration may be applied if required.

Pathological changes and detection of the poison in the corpse

From an examination of autopsy reports[21, 69, 116], it becomes apparent that histopathological studies contribute little of value in detecting the cause of death. No changes occur which could be considered as specific for fluoroacetate poisoning[31, 49, 63, 129]. For example, in the hospital records of an unpublished fatal case[118] made available to the author, the primary lesion which caused the patient's death was stated to be a diffuse, severe, degenerative process involving the heart, liver, kidney, brain and other organs; the finding of extensive early fatty degeneration and acute parenchymatous degeneration in these organs thus indicates the widespread yet uncharacteristic effect of the poison throughout the body. A few random observations may be mentioned however. Hicks[72] has found that, in rats, the testes of almost all males examined showed varying degrees of necrosis of the germinal cells of the seminiferous tubules, with characteristic confluent masses of dead or dying cells. Lesions occurred in the heart[72], caused by necrosis of muscle fibres with marked proliferation of mononuclear cells. Using rats, Kelemen, Ambrus and Ambrus[86] have observed pronounced histological changes in the ear following intraperitoneal injection of sodium fluoroacetate.

Certain biochemical changes might be developed as corroborative evidence of fluoroacetate poisoning. For example, there is a sevenfold increase of ammonium ion in the cerebrum of the dog[9], reaching a maximum after about 1 hour (at approximately the time at which epileptiform seizures develop); it is probable that a similar increase occurs in the brain of man. Again, fluoroacetate poisoning in rats causes a marked rise in blood-sugar levels[46], referred to by Engel and colleagues as 'sodium fluoroacetate diabetes'[40, 48]. However, the most characteristic and reliable biochemical effect is surely the massive increase in citrate in certain tissues, particularly in the kidney; such citrate accumulation has been observed in a very wide range of animals[133], and it seems almost certain that it occurs also in man (although this has not yet been verified in practice). Since citrate may be estimated reliably and without much difficulty[27, 117], such an effect could be considered

References p. 68

as *a priori* evidence of fluoroacetate poisoning. However, it is essential that such an estimation be carried out with all possible speed, since the citrate accumulation disappears rapidly after death, and is almost negligible after 1 hour[8]; this disappearance, probably due to the action of aconitase following diffusion out of the cell, might be retarded however by freezing the appropriate tissues in dry ice. The diagnostic value of citrate accumulation is rendered less specific if monoacetin or acetamide therapy has been used in an effort to save the life of the victim: both these compounds produce a rise in the citrate levels even in the absence of fluoroacetate[55].

It is possible that the unequivocal demonstration of organically-bound fluorine might be the most reliable proof of fluoroacetate poisoning; simple demonstration of the fluoride ion would of course not be adequate, since this exists naturally, particularly in organs of support or protection, such as bone, hair, nails and epidermis[103]. This is not the place for a survey of qualitative and quantitative analytical procedures, and the reader is referred to an excellent and comprehensive review of the subject[47]. In considering the small amounts likely to be encountered, the author favors the following procedure as being the most promising: (a) Estimation of free fluoride ion in the most appropriate tissue (kidney would probably be the most convenient and reliable); the recent method of Hall[66] provides a convenient means of determining microquantities, using paper chromatography. (b) Treatment of a small portion of the tissue to convert organically-bound fluorine to fluoride ion[47], followed by estimation of the total fluoride. The difference between (a) and (b) represents the quantity of organically-bound fluorine present in the tissue; this in a person not suffering from fluoroacetate poisoning should be negligible.

When the necessary specialized facilities are available, excellent and reliable detection and estimation may be achieved using nuclear magnetic resonance spectroscopy[41, 64, 146]. For this, about 10 mg of sample is normally required, which would have to be isolated by appropriate extraction procedures on the stomach, kidney, brain, body fluids, etc.; such a quantity is by no means unrealistic, since, in the only reported case in which estimation of the poison was attempted, 465 mg were calculated to be present in the corpse[69]. When only very small samples are available, simple identification of organically-bound fluorine may be achieved with as little as 0.5 mg of material. The technique can be used to differentiate directly between organically-bound fluorine and inorganic fluorides, hence no preliminary treatment of the sample is necessary; moreover, it is possible to distinguish between primary and secondary aliphatic fluorides, and therefore between fluoroacetate and fluorocitrate. It is likely that as further research progresses, the sensitivity and accuracy of detection using this new and valuable technique will improve still further.

As alternative procedures, the C-F bond may be identified by infra-red analysis[47, 54], but it is unlikely that this in itself would be sufficient for legal purposes. Finally, mention may be made of a report[14] in which it is claimed that microquantities of fluoroacetate itself may be determined in biological materials by conversion to fluoroacetylhydroxamic acid followed by two-dimensional paper chromatography and dye formation; however, this would not be diagnostic in cases where most of the ingested fluoroacetate had already been transformed to fluorocitrate *in vivo*.

In summary, the two most promising procedures for proving fluoroacetate poisoning involve: (a) citrate determinations; and (b) detection and estimation of organically-bound fluorine by chemical or by physical means. But even now it is improbable that a general practitioner would have the necessary facilities or specialized knowledge to make a correct diagnosis. Hence in the interests of safety, recourse must still be made to such time-honoured methods as detailed inventories and signed records of sales.

REPRESENTATIVE COMPOUNDS

In Table VI are shown some fluoroacetates; the list is not exhaustive, and is restricted to representative compounds, the toxicity of which has been reported. If several references to the same compound exist, usually the one giving toxicity is quoted. The LD_{50} figures

References p. 68

refer only to results obtained using *mice*. It is appreciated that there are disadvantages in limiting the record to this one species, particularly when the response is so remarkably varied between different species. However, the majority of work on fluoroacetates has been done with mice (presumably since they are the most readily available test animals); hence results on mice comprise the most nearly universal yardstick for a rough quantitative comparison of different compounds.

NOTE TO TABLES

The following arbitrary generalizations have been adopted in Table VI and all subsequent Tables to classify the LD_{50} figures (mg/kg) : $<$ 3.0 : very toxic; 3–25 : toxic; 25–75 : indefinite; $>$ 75 : non-toxic. If no LD_{50} figure is listed for a compound, either the toxicological conclusion has been reached using animals other than mice, or else no precise figure was quoted in the literature source. It should be emphasized that the LD_{50} figures for mice cannot be used for calculating a fatal dose for man, although it might be reasonable to conclude that the relative order of toxicity of the different compounds would be the same for both species. Routes of administration have not been included in the Table, since all apparently give rise to very similar toxicity results (p. 34). A few compounds which are not fluoroacetates *per se* have been included for comparison.

TABLE VI

Class	*Compound*	*Formula*
	(a) *Fluoroacetic acid and derivatives*	
Acid	Fluoroacetic acid	FCH_2COOH
	Difluoroacetic acid	$F_2CHCOOH$
	Trifluoroacetic acid	F_3CCOOH
Aldehyde	Fluoroacetaldehyde	FCH_2CHO
Salt	Sodium fluoroacetate	FCH_2COONa
	Triethyl-lead fluoroacetate	$FCH_2COOPbEt_3$
Ester	Methyl fluoroacetate	FCH_2COOCH_3
	Ethyl fluoroacetate	$FCH_2COOC_2H_5$
	n-Propyl fluoroacetate	$FCH_2COOCH_2CH_2CH_3$
	Isopropyl fluoroacetate	$FCH_2COOCH(CH_3)_2$
	Undecyl fluoroacetate	$FCH_2COO(CH_2)_{10}CH_3$
	Lauryl fluoroacetate	$FCH_2COO(CH_2)_{11}CH_3$
	Allyl fluoroacetate	$FCH_2COOCH_2CH{=}CH_2$
	2-Chloroethyl fluoroacetate	$FCH_2COOCH_2CH_2Cl$
	2-Fluoroethyl fluoroacetate	$FCH_2COOCH_2CH_2F$
	Phenyl fluoroacetate	$FCH_2COOC_6H_5$
	Fluoroacetyl salicylic acid	$FCH_2COOC_6H_4COOH$
	Fluoroacetylcholine bromide	$FCH_2COOCH_2CH_2NMe_3Br$
	Methyl fluorothiolacetate	FCH_2COSCH_3
	Ethyl fluorothiolacetate	$FCH_2COSC_2H_5$
	2-Chloroethyl fluorothiol-acetate	$FCH_2COSCH_2CH_2Cl$
	Phenyl fluorothiolacetate	$FCH_2COSC_6H_5$
	Methylene bis-fluoroacetate	$(FCH_2COO)_2CH_2$
	Ethylene glycol bis-fluoroacetate	$(FCH_2COOCH_2)_2$
	Thiodiglycol bis-fluoroacetate	$(FCH_2COOCH_2CH_2)_2S$
	Cholesteryl fluoroacetate	$FCH_2COOC_{27}H_{45}$
	Methyl 2-fluoropropionate**	$CH_3CHFCOOCH_3$
	Methyl 2-fluoro-isobutyrate**	$(CH_3)_2CFCOOCH_3$
	Methyl difluoroacetate	$F_2CHCOOCH_3$
	Methyl trifluoroacetate	$F_3CCOOCH_3$
	Methyl chlorofluoroacetate	$FCHClCOOCH_3$
	Methyl dichlorofluoroacetate	$FCCl_2COOCH_3$

LD₅₀ (mice) mg/kg*	*Conclusion**	*Boiling point*	*Reference*
6–10	toxic	167–168.5°	31, 125, 149
—	non-toxic	134–135°	63
—	non-toxic	72–72.5°	63
6	toxic	64–65°	89, 152
6–10	toxic	—	31, 149
15	toxic	m.p. 180.5° (decomp.)	149
6–10	toxic	104.5°	31, 125, 149
6–10	toxic	115–117°	125, 126, 149
6–10	toxic	135–137°	149
6–10	toxic	148.5°	149
60	indefinite	97–97.5°/0.5 mm	121
30	indefinite	146–150°/4 mm	121
6	toxic	136–137°	150
—	toxic	178–179°	150
—	toxic	90.5–91°/58 mm	150
6–10	toxic	m.p. 63.5–64°	149
15	toxic	m.p. 131.6°	149
—	toxic	m.p. 124°	17, 61, 147
6–10	toxic	119–120°	119, 121
6–10	toxic	135–136°	119, 121
17.5	toxic	104–105°/33 mm	150
80	non-toxic	132°/18 mm	150
10	toxic	m.p. 57°	149
—	toxic	140–141°/11 mm	149
—	toxic	161–162°/3 mm	63
—	non-toxic	m.p. 144.5°	63, 149
—	non-toxic	106.5–108.5°	149
—	non-toxic	108–109°	149
—	non-toxic	85–86°	62, 63
—	non-toxic	43–43.5°	63
—	non-toxic	110–111°	63
—	non-toxic	116–116.5°	63

References p. 68

TABLE VI *(continued)*

Class	*Compound*	*Formula*
	Ethyl fluoroformate	$FCOOC_2H_5$
	cf. Methyl chloroacetate	$ClCH_2COOCH_3$
Amide	Fluoroacetamide	FCH_2CONH_2
	N-Methylfluoroacetamide	$FCH_2CONHCH_3$
	N-Nitroso-N-methylfluoro-acetamide	$FCH_2CON(NO)CH_3$
	N-2-Chloroethylfluoro-acetamide	$FCH_2CONHCH_2CH_2Cl$
	Ethyl fluoroacetamidoacetate	$FCH_2CONHCH_2COOC_2H_5$
Acid halide	Fluoroacetyl chloride	FCH_2COCl
	Fluoroacetyl bromide	FCH_2COBr
	Fluoroacetyl fluoride	FCH_2COF
	Acetyl fluoride	CH_3COF
	Chloroacetyl fluoride	$ClCH_2COF$
Anhydride	Fluoroacetic anhydride	$(FCH_2CO)_2O$
Nitrile	Fluoroacetonitrile	FCH_2CN
Miscella-neous	Fluoroacetimino ethyl ether hydrochloride	$FCH_2C(OEt){=}NH_2Cl$
	Fluoroacetimino 2-fluoroethyl ether hydrochloride	$FCH_2C(OCH_2CH_2F){=}NH_2Cl$
	Fluoroacetamidine hydro-chloride	$FCH_2C(NH_2){=}NH_2Cl$
	Monofluoroacetone	FCH_2COCH_3
	ω-Fluoroacetophenone	$FCH_2COC_6H_5$
	(b) *2-Fluoroethanol and derivatives*	
Alcohol	2-Fluoroethanol	FCH_2CH_2OH
Ester	2-Fluoroethyl acetate	$FCH_2CH_2OCOCH_3$
	2-Fluoroethyl caproate	$FCH_2CH_2OCO(CH_2)_4CH_3$
	2-Fluoroethyl laurate	$FCH_2CH_2OCO(CH_2)_{10}CH_3$
	2-Fluoroethyl oleate	$FCH_2CH_2OCOC_{17}H_{33}$
	2-Fluoroethyl benzoate	$FCH_2CH_2OCOC_6H_5$
	2-Fluoroethyl chloroacetate	$FCH_2CH_2OCOCH_2Cl$
	2-Fluoroethyl aminoacetate hydrochloride	$FCH_2CH_2OCOCH_2NH_3Cl$

FLUOROACETATES

*LD₅₀** *(mice) mg/kg*	*Conclusion**	*Boiling point*	*Reference*
—	non-toxic	55.5°	149
—	non-toxic	131.5°	149
6–10	toxic	m.p. 108°	22
6–10	toxic	m.p. 64°	22
6–10	toxic	b.p. 84°/14 mm	22
15	toxic	m.p. 65°	22
20	toxic	m.p. 50–50.5°	153
6–10	toxic	71.5–73°	149
—	toxic	95–96°	63, 122
6–10	toxic	50.5–51°	149
—	non-toxic	20.5°	149
—	non-toxic	73–76°	149
6–10	toxic	88–89°/12 mm	149
25	indefinite	79–80°	120
6–10	toxic	solid	22
—	toxic	solid	22
6–10	toxic	m.p. 222–224° (decomp.)	22
—	non-toxic	78–79°	57, 126
> 225	non-toxic	90–91°/12 mm	13, 63
10	toxic	100–101°	124, 152
18.6	toxic	116–117.5°	109
> 96	non-toxic	91–93°/24 mm	121
32	indefinite	153–155°/13 mm	121
200	non-toxic	169–170°/1 mm	121
38	indefinite	112–114°/10 mm	109
6–10	toxic	178°	150
10	toxic	m.p. 150–150.5°	153

TABLE VI *(continued)*

Class	*Compound*	*Formula*
	2-Fluoroethyl betaine hydrochloride	$FCH_2CH_2OCOCH_2NMe_3Cl$
	2-Fluoroethyl methane-sulphonate	$FCH_2CH_2OSO_2CH_3$
	2-Fluoroethyl *p*-toluene-sulphonate	$FCH_2CH_2OSO_2C_6H_4CH_3$
	2-Fluoroethyl ethyl carbonate	$FCH_2CH_2OCOOC_2H_5$
	Bis-2-fluoroethyl carbonate	$(FCH_2CH_2O)_2CO$
	Bis-2-fluoroethyl sulphate	$(FCH_2CH_2O)_2SO_2$
	Bis-2-fluoroethyl sulphite	$(FCH_2CH_2O_2SO$
	2-Fluoroethyl chlorosulphonate	$FCH_2CH_2OSO_2Cl$
	Bis-2-fluoroethyl phosphoro-fluoridate	$(FCH_2CH_2O)_2POF$
Halide†	2-Fluoroethyl chloride	FCH_2CH_2Cl
	2-Fluoroethyl bromide	FCH_2CH_2Br
	2-Fluoroethyl iodide	FCH_2CH_2I
Ammonium salt	2-Fluoroethyl-trimethyl-ammonium bromide	$FCH_2CH_2NMe_3Br$
	2-Fluoroethyl-pyridinium bromide	$FCH_2CH_2NC_5H_5Br$
	3-Carbethoxy-N-2-fluoroethyl-pyridinium bromide	$FCH_2CH_2NC_5H_4(COOEt)Br$
Urethane	2-Fluoroethyl carbamate	$FCH_2CH_2OCONH_2$
	2-Fluoroethyl N-methyl-carbamate	$FCH_2CH_2OCONHCH_3$
	2-Fluoroethyl N,N-dimethyl-carbamate	$FCH_2CH_2OCON(CH_3)_2$
Miscellaneous	Diethyl 2-fluoroethyl-phosphonate	$FCH_2CH_2PO(OEt)_2$
	2-Fluoro-2′-hydroxydiethyl ether	$FCH_2CH_2OCH_2CH_2OH$
	2-Fluoroethyl *β*-naphthyl ether	$FCH_2CH_2OC_{10}H_7$
	Bis-2-fluoroethyl methylal	$(FCH_2CH_2O)_2CH_2$
	Bis-2-(2′-fluoroethoxy)ethyl methylal	$(FCH_2CH_2OCH_2CH_2O)CH_2$

* See qualifying remarks (p. 61). ** See footnote (p. 82). † See footnote (p. 120).

LD_{50} (mice) mg/kg*	*Conclusion**	*Boiling point*	*Reference*
45	indefinite	m.p. 122°	153
ca. 100	non-toxic	118–119°/12 mm	109
> 100	non-toxic	138.5–140°/1 mm	109
6–10	toxic	56–57°/14 mm	24
10–15	toxic	89–90°/14 mm	157
—	indefinite	145°/18 mm	152
10–15	toxic	108°/17 mm	157
—	indefinite	80°/18 mm	152
—	indefinite	125–127°/13 mm	30
—	non-toxic	53.5–54°	123, 152
—	non-toxic	70–71°	123, 152
28	indefinite	89–91°	123
300	non-toxic	m.p. 244°	153
300	non-toxic	m.p. 180°	153
200	non-toxic	m.p. 86-88°	153
—	toxic	105°/12 mm	157
—	toxic	92°/13 mm	157
—	toxic	75–80°/5 mm	114
—	non-toxic	74–75°/11 mm	151
—	toxic	75°/12 mm	157
60	indefinite	m.p. 49.5–50°	152
—	toxic	155°	157
—	toxic	149°/13 mm	157

References p. 68

REFERENCES

1. Anon. (1940) Improvements in or relating to the production of substituted acetic acids. *British Patent 527, 644*, October 14, 1940.
2. Anon. (1948) *Instructions for using sodium fluoroacetate (compound 1080) as a rodent poison.* National Research Council, Chemical-Biological Coordination Center, Washington, D.C. (Revised October, 1948).
3. Anon. (1956) *Clinical memoranda on economic poisons.* Technical Development Laboratories, Communicable Disease Center, U.S. Public Health Service, Savannah, Georgia, Revised April 1, 1956, pp. 52–55.
4. BACON, J. C., BRADLEY, C. W., HOEGBERG, E. I., TARRANT, P., and CASSADAY, J. T. (1948) Some amides and esters of fluoroacetic acid. *J. Am. Chem. Soc., 70:* 2653.
5. BADENHUIZEN, N. P., and SLINGER, J. (1954) Detection of monofluoroacetic acid in Gifblaar, *Dichapetalum cymosum.* I. *S. African J. Sci., 50:* 269.
6. BALDWIN, H. B. (1899) The toxic action of sodium fluoride. *J. Am. Chem. Soc., 21:* 517.
7. BARTLETT, G. R., and BARRON, E. S. G. (1947) The effect of fluoroacetate on enzymes and on tissue metabolism. Its use for the study of the oxidative pathway of pyruvate metabolism. *J. Biol. Chem., 170:* 67.
8. BEAULIEU, M. M., and DALLEMAGNE, M. J. (1953) Modifications du taux d'acide citrique du foie et des reins sous l'influence du fluoracétate sodique. *Bull. soc. chim. biol., 35:* 969.
9. BENITEZ, D., PSCHEIDT, G. R., and STONE, W. E. (1954) Formation of ammonium ion in the cerebrum in fluoroacetate poisoning. *Am. J. Physiol., 176:* 488.
10. BERGMANN, E. D., and BLANK, I. (1953) Studies on organic fluorine compounds. Part I. Some esters of monofluoroacetic acid and related compounds. *J. Chem. Soc., 1953:* 3786.
11. BERGMANN, E. D., and SCHWARCZ, J. (1956) Organic fluorine compounds. Part VII. The Perkin and similar reactions with fluoroacetic acid. *J. Chem. Soc., 1956:* 1524.
12. BERGMANN, E. D., and SZINAI, S. (1956) Organic fluorine com-

pounds. Part VI. The enolates of alkyl fluoroacetates. *J. Chem. Soc.*, *1956:* 1521.

13. BERGMANN, F., and KALMUS, A. (1954) Synthesis and properties of ω-fluoroacetophenone. *J. Am. Chem. Soc.*, *76:* 4137.
14. BERGMANN, F., and SEGAL, R. (1956) The separation and determination of microquantities of lower aliphatic acids, including fluoroacetic acid. *Biochem. J.*, *62:* 542.
15. BERGMANN, F., and SHIMONI, A. (1953) The enzymic hydrolysis of alkyl fluoroacetates and related compounds. *Biochem. J.*, *55:* 50.
16. BERGMANN, M., and FRUTON, J. S., Personal communication to CHENOWETH, M. B. (1949) *J. Pharmacol. Exptl. Therap.*, *II*, *97:* 383. *Pharmacol. Revs.*, *1:* 383.
17. BLOHM, T. R. (1951) Fluoroacetylcholine bromide and some other choline ester salts. *J. Am. Chem. Soc.*, *73:* 5445.
18. BOYARSKY, L. L., ROSENBLATT, A. D., POSTEL, S., and GERARD, R. W. (1949) Action of methyl fluoroacetate on respiration and potential of nerve. *Am. J. Physiol.*, *157:* 291.
19. BRADY, R. O. (1955) Fluoroacetyl coenzyme A. *J. Biol. Chem.*, *217:* 213.
20. BRAUNSTEIN, A. E., and AZARKH, R. M. (1957) The effect of fluorocitrate on the synthesis of amino acids from ammonia and α-keto acids in rat liver homogenate. *Arch. Biochem. Biophys.*, *69:* 634.
21. BROCKMANN, J. L., MCDOWELL, A. V., and LEEDS, W. G. (1955) Fatal poisoning with sodium fluoroacetate. Report of a case. *J. Am. Med. Assoc.*, *159:* 1529.
22. BUCKLE, F. J., HEAP, R., and SAUNDERS, B. C. (1949) Toxic fluorine compounds containing the C-F link. Part III. Fluoroacetamide and related compounds. *J. Chem. Soc.*, *1949:* 912.
23. BUCKLE, F. J., PATTISON, F. L. M., and SAUNDERS, B. C. (1949) Toxic fluorine compounds containing the C-F link. Part VI. ω-Fluorocarboxylic acids and derivatives. *J. Chem. Soc.*, *1949:* 1471.
24. BUCKLE, F. J., and SAUNDERS, B. C. (1949) Toxic fluorine compounds containing the C-F link. Part VIII. ω-Fluorocarboxylic acids and derivatives containing an oxygen atom as a chain member. *J. Chem. Soc.*, *1949:* 2774.

25. BUFFA, P. Unpublished results quoted by PETERS, R. A. (1957) *Advances in Enzymology and Related Subjects of Biochemistry*, Vol. XVIII, Interscience Publishers, Inc., New York, p. 138.

26. BUFFA, P., and PETERS, R. A. (1949) Formation of citrate *in vivo* induced by fluoroacetate poisoning. *Nature*, *163:* 914.

27. BUFFA, P., and PETERS, R. A. (1950) The *in vivo* formation of citrate induced by fluoroacetate and its significance. *J. Physiol.* (London), *110:* 488.

28. BUFFA, P., PETERS, R. A., and WAKELIN, R. W. (1951) Biochemistry of fluoroacetate poisoning. Isolation of an active tricarboxylic acid fraction from poisoned kidney homogenates. *Biochem. J.*, *48:* 467.

29. CHAPMAN, C., and PHILLIPS, M. A. (1956) Fluoroacetamide, and potassium and sodium fluoroacetates therefrom. *British Patent 757, 610*, September 19, 1956.

30. CHAPMAN, N. B., and SAUNDERS, B. C. (1948) Esters containing phosphorus. Part VI. Preparation of esters of fluorophosphonic acid by means of phosphorus oxydichlorofluoride. *J. Chem. Soc.*, *1948:* 1010.

31. CHENOWETH, M. B. (1949) Monofluoroacetic acid and related compounds. *J. Pharmacol. Exptl. Therap.*, *II*, *97:* 383; *Pharmacol. Revs.*, *1:* 383.

32. CHENOWETH, M. B. (1949) Unpublished results.

33. CHENOWETH, M. B. (1950) Personal communication, September 18, 1950.

34. CHENOWETH, M. B. (1958) Personal communication, September 18, 1958.

35. CHENOWETH, M. B., and GILMAN, A. (1946) Studies on the pharmacology of fluoroacetate. I. Species response to fluoroacetate. *J. Pharmacol. Exptl. Therap.*, *87:* 90.

36. CHENOWETH, M. B., and GILMAN, A. (1947) Studies on the pharmacology of fluoroacetate. II. Action on the heart. *Bull. U.S. Army Med. Dept.*, *7:* 687.

37. CHENOWETH, M. B., KANDEL, A., JOHNSON, L. B., and BENNETT, D. R. (1951) Factors influencing fluoroacetate poisoning. Practical treatment with glycerol monoacetate. *J. Pharmacol. Exptl. Therap.*, *102:* 31.

38. CHENOWETH, M. B., SCOTT, E. B., and SEKI, S. L. (1949) Prevention of fluoroacetate poisoning by acetate donors. *Federation Proc.*, *8:* 280.
39. CHENOWETH, M. B., and ST. JOHN, E. F. (1947) Studies on the pharmacology of fluoroacetate. III. Effects on the central nervous systems of dogs and rabbits. *J. Pharmacol. Exptl. Therap.*, *90:* 76.
40. COLE, B. T., ENGEL, F. L., and FREDERICKS, J. (1955) Sodium fluoroacetate diabetes: correlations between glycemia, ketonemia, and tissue citrate levels. *Endocrinology*, *56:* 675.
41. Conference speakers (GUTOWSKY, H. S., and NACHOD, F. C., Co-Chairmen) (1958) Nuclear magnetic resonance. *Annals N.Y. Acad. Sci.*, *70:* 763.
42. DEONIER, C. C., JONES, H. A., and INCHO, H. H. (1946) Organic compounds effective against *Anopheles quadrimaculatus*. Laboratory tests. *J. Econ. Entomol.*, *39:* 459.
43. DIEKE, S. H., and RICHTER, C. P. (1946) Comparative assays of rodenticides on wild Norway rats. I. Toxicity. *U.S. Public Health Repts.*, *61:* 672.
44. DIETRICH, L. S., and SHAPIRO, D. M. (1956) Fluoroacetate and fluorocitrate antagonism of tumor growth. Effect of these compounds on citrate metabolism in normal and neoplastic tissue. *Cancer Research*, *16:* 585.
45. ELLIOTT, W. B., and KALNITSKY, G. (1950) A mechanism for fluoroacetate inhibition. *J. Biol. Chem.*, *186:* 487.
46. ELLIOTT, W. B., and PHILLIPS, A. H. (1954) Effect of fluoroacetate on glucose metabolism *in vivo*. *Arch. Biochem. Biophys.*, *49:* 389.
47. ELVING, P. J., HORTON, C. A., and WILLARD, H. H. (1954) Analytical chemistry of fluorine and fluorine-containing compounds. In *Fluorine Chemistry* Vol. II (Editor, J. H. SIMONS), Academic Press Inc., New York, p. 51.
48. ENGEL, F. L., HEWSON, K., and COLE, B. T. (1954) Carbohydrate and ketone body metabolism in the sodium fluoroacetate-poisoned rat. 'Sodium fluoroacetate diabetes'. *Am. J. Physiol.*, *179:* 325.
49. FOSS, G. L. (1948) The toxicology and pharmacology of methyl fluoroacetate (MFA) in animals, with some notes on experimental therapy. *Brit. J. Pharmacol.*, *3:* 118.
50. FREEMAN, S., and ELLIOTT, J. R. (1956) The effect of fluoroacetate

upon the plasma citrate response to parathyroidectomy and nephrectomy. *Endocrinology*, *59:* 190.

51. Gaines, T. B., Sumerford, W. T., and Hayes, W. J. (1950) The non-toxicity of urine from rats poisoned with 1080. *Pest Control*, *18*, *No. 6:* 12.
52. Gajdusek, D. C., and Luther, G. (1950) Fluoroacetate poisoning. A review and report of a case. *Am. J. Diseases Children*, *79:* 310.
53. Gal, E. M., Peters, R. A., and Wakelin, R. W. (1956) Some effects of synthetic fluoro compounds on the metabolism of acetate and citrate. *Biochem. J.*, *64:* 161.
54. Gillieson, A. H. C. P., and Newcomb, R. A. (1951) The application of the hollow cathode source to spectrochemical analysis. Part II. The micro-determination of organically bound fluorine. *A.E.R.E. (Harwell) Report C/R 764*, Ministry of Supply, Great Britain.
55. Gitter, S. (1956) The influence of acetamide on citrate accumulation after fluoroacetate poisoning. *Biochem. J.*, *63:* 182.
56. Gitter, S., Blank, I., and Bergmann, E. D. (1953) Studies on organic fluorine compounds. I. The influence of acetamide on fluoroacetate poisoning. *Koninkl. Ned. Akad. Wetenschap. Proc. Ser. C.*, *56:* 423.
57. Gitter, S., Blank, I., and Bergmann, E. D. (1953) Studies on organic fluorine compounds. II. Toxicology of higher alkyl fluoroacetates. *Koninkl. Ned. Akad. Wetenschap. Proc. Ser. C.*, *56:* 427.
58. Gratch, I., Purlia, P. L., and Martin, M. L. (1949) Effect of sodium fluoroacetate (1080) in poisoned rats on plague diagnosis procedures. Preliminary report. *U.S. Public Health Repts.*, *64:* 339.
59. Gryszkiewicz-Trochimowski, E. (1947) Recherches sur les composés organiques fluorés dans la série aliphatique. III. Sur quelques alcools aliphatiques fluorés. *Rec. trav. chim.*, *66:* 427.
60. Gryszkiewicz-Trochimowski, E. (1947) Recherches sur les composés organiques fluorés dans la série aliphatique. IV. Synthèse des acides β-fluoro-propionique et γ-fluoro-butyrique. *Rec. trav. chim.*, *66:* 430.
61. Gryszkiewicz-Trochimowski, E., Gryszkiewicz-Trochimowski, O., and Lévy, R. (1953) Recherches sur les composés organiques

fluorés dans la série aliphatique. VII. Note sur la préparation de l'acide monofluoroacétique et sur plusieurs de ses dérivés. *Bull. soc. chim. France, 1953:* 462.

62. GRYSZKIEWICZ-TROCHIMOWSKI, E., SPORZYNSKI, A., and WNUK, J. (1947) Recherches sur les composés organiques fluorés dans la série aliphatique. I. Méthode genérale de préparation des composés organiques fluorés. *Rec. trav. chim., 66:* 413.

63. GRYSZKIEWICZ-TROCHIMOWSKI, E., SPORZYNSKI, A., and WNUK, J. (1947) Recherches sur les composés organiques fluorés dans la série aliphatique. II. Sur les dérivés des acides mono-, di- et tri-fluoro-acétiques. *Rec. trav. chim., 66:* 419.

64. GUTOWSKY, H. S. (1956) Analytical applications of nuclear magnetic resonance. In *Physical Methods in Chemical Analysis* Vol. III, (Editor, W. G. BERL), Academic Press, Inc., New York, p. 303.

65. HAGAN, E. C., RAMSEY, L. L., and WOODARD, G. (1950) Absorption, distribution, and excretion of sodium fluoroacetate (1080) in rats. *J. Pharmacol. Exptl. Therap., 99:* 432.

66. HALL, R. J. (1957) The use of paper chromatography for the detection and determination of microgram amounts of inorganic fluoride. *Analyst, 82:* 663.

67. HARRISON, P. K. (1949) *New compounds as insecticides against the sweet-potato weevil.* Bur. Entomol. and Plant Quarantine E 770, 5 pp.

68. HARRISSON, J. W. E., AMBRUS, J. L., and AMBRUS, C. M. (1952) Fluoroacetate (1080) poisoning. *Ind. Med. and Surg., 21:* 440.

69. HARRISSON, J. W. E., AMBRUS, J. L., AMBRUS, C. M., REES, E. W., PETERS, R. H., REESE, L. C., and BAKER, T. (1952) Acute poisoning with sodium fluoroacetate (Compound 1080). *J. Am. Med. Assoc., 149:* 1520.

70. HASTINGS, A. B., PETERS, R. A., and WAKELIN, R. W. (1955) The effect of subarachnoid injection of fluorocitrate into cerebrospinal fluid of rabbits. *Quart. J. Exptl. Physiol., 40:* 258.

71. HENDERSHOT, L. C., and CHENOWETH, M. B. (1955) Fluoroacetate and fluorobutyrate convulsions in the isolated cerebral cortex of the dog. *J. Pharmacol. Exptl. Therap., 113:* 160.

72. HICKS, S. P. (1950) Brain metabolism *in vivo.* I. The distribution of lesions caused by cyanide poisoning, insulin hypoglycemia,

asphyxia in nitrogen and fluoroacetate poisoning in rats. *Arch. Pathol., 49:* 111.

73. HOFFMANN, F. W. (1948) Preparation of aliphatic fluorides. *J. Am. Chem. Soc., 70:* 2596.
74. HOFFMANN, F. W. (1949) Personal communication, October 14, 1949.
75. HORSFALL, J. L. (1946) 2-Ethylhexyl fluoroacetate. *U.S. Patent 2,409,859*, October 22, 1946.
76. HUGHES, G. M. K., and SAUNDERS, B. C. (1954) Studies in peroxidase action. Part IX. Reactions involving the rupture of the C-F, C-Br, and C-I links in aromatic amines. *J. Chem. Soc., 1954:* 4630.
77. HUTCHENS, J. O., WAGNER, H., PODOLSKY, B., and MCMAHON, T. M. (1949) The effect of ethanol and various metabolites on fluoroacetate poisoning. *J. Pharmacol. Exptl. Therap., 95:* 62.
78. JENKINS, R. L., and KOEHLER, H. C. (1948) Making 1080 safe. A case study in the safe manufacture and distribution of a hazardous chemical. *Chem. Ind., 62:* 232.
79. JUDAH, J. D., and REES, K. R. (1954) A note on the action of fluoroacetate. *Biochem. J., 57:* 374.
80. KACSER, H. (1955) The stereochemical relations of the four fluorocitrate isomers. *Disc. Faraday Soc., 20:* 283.
81. KALCKAR, H. M. (1952) Metabolic enzymes of mitochondria. *Acta. Med. Scand., 142, Suppl. 266:* 615.
82. KALMBACH, E. R. (1945) 'Ten-eighty', a war-produced rodenticide. *Science, 102:* 232.
83. KALNITSKY, G., and BARRON, E. S. G. (1948) The inhibition by fluoroacetate and fluorobutyrate of fatty acid and glucose oxidation produced by kidney homogenates. *Arch. Biochem., 19:* 75.
84. KANDEL, A., and CHENOWETH, M. B. (1952) Metabolic disturbances produced by some fluoro-fatty acids: relation to the pharmacologic activity of these compounds. *J. Pharmacol. Exptl. Therap., 104:* 234.
85. KANDEL, A., and CHENOWETH, M. B. (1952) Tolerance to fluoroacetate and fluorobutyrate in rats. *J. Pharmacol. Exptl. Therap., 104:* 248.
86. KELEMEN, G., AMBRUS, J. L., and AMBRUS, C. M. (1955) Fluoro-

acetate poisoning and the hearing organ. Experimental study. *Ann. Otol., Rhinol. & Laryngol., 64:* 1046.

87. KHARASCH, M. S., JENSEN, E. V., and URRY, W. H. (1945) Reactions of atoms and free radicals in solution. VI. Decomposition of diacetyl and other peroxides in aliphatic acids and substituted aliphatic esters. *J. Org. Chem., 10:* 386.
88. KING, J. E., and PENFOUND, W. T. (1946) Effects of new herbicides on fish. *Science, 103:* 487.
89. KITANO, H., and FUKUI, K. (1955) Preparation of some fluoroethanol derivatives. *J. Chem. Soc. Japan, Ind. Chem. Sect., 58:* 355.
90. KNUNYANTS, I. L., KIL'DISHEVA, O. V., and BYKHOVSKAYA, E. (1949) The reaction of aliphatic oxides with hydrogen fluoride. II. *J. Gen. Chem. U.S.S.R., 19:* 93 [Engl. translation].
91. KNUNYANTS, I. L., KIL'DISHEVA, O. V., and PETROV, I. P. (1949) The reaction of aliphatic oxides with hydrogen fluoride. I. *J. Gen. Chem. U.S.S.R., 19:* 87 [Engl. translation].
92. KUMASAWA, N. (1954) Poisonous effects of fluorine compounds on mouse and field mouse. Part I. *Ann. Rept. Takamine Lab., 6:* 141.
93. LARNER, J. (1950) Toxicological and metabolic effects of fluorine-containing compounds. *Ind. Med. and Surg., 19:* 535.
94. LIÉBECQ, C., and PETERS, R. A. (1948) The inhibitory effect of fluoroacetate and the tricarboxylic cycle. *Proc. Physiol. Soc., June 26, 1948, J. Physiol., Vol. 108.*
95. LIÉBECQ, C., and PETERS, R. A. (1949) The toxicity of fluoroacetate and the tricarboxylic acid cycle. *Biochim. Biophys. Acta, 3:* 215.
96. LINDENBAUM, A., WHITE, M. R., and SCHUBERT, J. (1951) Citrate formation *in vivo* induced by non-lethal doses of fluoroacetate. *J. Biol. Chem., 190:* 585.
97. LOTSPEICH, W. D., PETERS, R. A., and WILSON, T. H. (1952) The inhibition of aconitase by 'inhibitor fractions' isolated from tissues poisoned with fluoroacetate. *Biochem. J., 51:* 20.
98. MACCHIAVELLO, A. (1946) Plague control with DDT and '1080'. Results achieved in a plague epidemic at Tumbes, Peru, 1945. *Am. J. Public Health, 36:* 842.
99. MARAIS, J. S. C. (1943) The isolation of the toxic principle 'potas-

sium cymonate' from 'Gifblaar', *Dichapetalum cymosum* (Hook) Engl. *Onderstepoort J. Vet. Sci. Animal Ind., 18:* 203.

100. MARAIS, J. S. C. (1944) Monofluoroacetic acid, the toxic principle of 'Gifblaar', *Dichapetalum cymosum* (Hook) Engl. *Onderstepoort J. Vet. Sci. Animal Ind., 20:* 67.

101. MARCUS, A., and ELLIOTT, W. B. (1956) Enzymic reactions of fluoroacetate and fluoroacetyl coenzyme A. *J. Biol. Chem., 218:* 823.

102. MARTIUS, C. (1949) Über die Unterbrechung des Citronensäure-Cyklus durch Fluoressigsäure. *Ann., 561:* 227.

103. MATUURA, S., KOKUBU, N., WATANABE, S., and SAMESHIMA, Y. (1955) Fluorine content of animals. *Mem. Fac. Sci., Kyushu Univ. Ser. C, 2:* 81.

104. MCBEE, E. T., and CHRISTMAN, D. L. (1955) Infra-red spectra of halogenated acetic esters. *J. Am. Chem. Soc., 77:* 755.

105. MCCOMBIE, H., and SAUNDERS, B. C. (1946) Fluoroacetates and allied compounds. *Nature, 158:* 382.

106. MCCOMBIE, H., and SAUNDERS, B. C. (1947) Toxic organo-lead compounds. *Nature, 159:* 491.

107. MILLER, R. F., and PHILLIPS, P. H. (1955) Effect of feeding fluoroacetate to the rat. *Proc. Soc. Exptl. Biol. Med., 89:* 411.

108. MILLER, W. T., and PROBER, M. (1948) The vapor phase fluorination of acetyl fluoride. *J. Am. Chem. Soc., 70:* 2602.

109. MILLINGTON, J. E., and PATTISON, F. L. M. (1956) Toxic fluorine compounds. XII. Esters of ω-fluoroalcohols. *Can. J. Chem., 34:* 1532.

110. MORRISON, J. F., and PETERS, R. A. (1954) Biochemistry of fluoroacetate poisoning: the effect of fluorocitrate on purified aconitase. *Biochem. J., 58:* 473.

111. MORRISON, J. L. (1946) Toxicity of certain halogen-substituted aliphatic acids for white mice. *J. Pharmacol. Exptl. Therap., 86:* 336.

112. NOZU, R., KITANO, H., and OSAKA, T. (1955) Reaction of haloacetic acid esters with potassium fluoride. *J. Chem. Soc. Japan, Ind. Chem. Sect., 58:* 12.

113. OLAH, G., and PAVLATH, A. (1953) Synthesis of organic fluorine compounds. II. The preparation of 2-fluoroethanol. *Acta Chim. Acad. Sci. Hung., 3:* 199.

114. OLIVERIO, V. T., and SAWICKI, E. (1955) Some fluorine derivatives of urethan. *J. Org. Chem., 20:* 363.

115. ORD, M. G., and STOCKEN, L. A. (1953) Sex differences in effect of radiation and fluoroacetate poisoning on liver metabolism. *Proc. Soc. Exptl. Biol. Med., 83:* 695.

116. PARKER, E. F. (1947) *Report of a case at the Public Hospital, Moline, Illinois to the National Pest Control Association*, 250 West Jersey Street, Elizabeth, N. J., October 22, 1947.

117. PARKER, J. M., and WALKER, I. G. (1957) A toxicological and biochemical study of ω-fluoro compounds. *Can. J. Biochem. Physiol., 35:* 407.

118. *Parkland Memorial Hospital*, Medical Record Library, 5201 Harry Hines Boulevard, Dallas, Texas (1947) *Clinical record of J.B.M.* (March 21, 1947) kindly supplied at the request of Dr. J. W. BASS, Director of Public Health, Dallas, Texas.

119. PATTISON, F. L. M. (1954) Toxic fluorine compounds. II. *Nature, 174:* 737.

120. PATTISON, F. L. M., COTT, W. J., HOWELL, W. C., and WHITE, R. W. (1956) Toxic fluorine compounds. V. ω-Fluoronitriles and ω-fluoro-ω′-nitroalkanes. *J. Am. Chem. Soc., 78:* 3484.

121. PATTISON, F. L. M., FRASER, R. R., MIDDLETON, E. J., SCHNEIDER, J. C., and STOTHERS, J. B. (1956) Esters of fluoroacetic acid and of 2-fluoroethanol. *Can. J. Technol., 34:* 21.

122. PATTISON, F. L. M., FRASER, R. R., O'NEILL, G. J., and WILSHIRE, J. F. K. (1956) Toxic fluorine compounds. X. ω-Fluorocarboxylic acid chlorides, anhydrides, amides and anilides. *J. Org. Chem., 21:* 887.

123. PATTISON, F. L. M., and HOWELL, W. C. (1956) Toxic fluorine compounds. IV. ω-Fluoroalkyl halides. *J. Org. Chem., 21:* 748.

124. PATTISON, F. L. M., HOWELL, W. C., MCNAMARA, A. J., SCHNEIDER, J. C., and WALKER, J. F. (1956) Toxic fluorine compounds. III. ω-Fluoroalcohols. *J. Org. Chem., 21:* 739.

125. PATTISON, F. L. M., HUNT, S. B. D., and STOTHERS, J. B. (1956) Toxic fluorine compounds. IX. ω-Fluorocarboxylic esters and acids. *J. Org. Chem., 21:* 883.

126. PATTISON, F. L. M., and MILLINGTON, J. E. (1956) The preparation and some cleavage reactions of alkyl and substituted alkyl methane-

sulphonates. The synthesis of fluorides, iodides, and thiocyanates. *Can. J. Chem., 34:* 757.

127. PAULING, L. (1939) *The nature of the chemical bond,* Cornell University Press, Ithaca, N.Y.

128. PAULING, L., and HUGGINS, M. L. (1934) Covalent radii of atoms and interatomic distances in crystals containing electron-pair bonds. *Z. Krist., 87:* 205.

129. PETERS, R. A. (1952) Lethal synthesis. [Croonian lecture.] *Proc. Roy. Soc. (London), B, 139:* 143.

130. PETERS, R. A. (1953) Significance of biochemical lesions in the pyruvate oxidase system. *Brit. Med. Bull., 9:* 116.

131. PETERS, R. A. (1955) Biochemistry of some toxic agents. II. Some recent work in the field of fluoroacetate compounds. [Dohme lectures] *Bull. Johns Hopkins Hospital, 97:* 21.

132. PETERS, R. A. (1955) The action of fluorocitric acid on aconitase. *Discussions Faraday Soc., No. 20:* 189.

133. PETERS, R. A. (1957) Mechanism of the toxicity of the active constituent of *Dichapetalum cymosum* and related compounds. *Advances in Enzymology and Related Subjects of Biochemistry.* Vol. XVIII, Interscience Publishers, Inc., New York, p. 113.

134. PETERS, R. A., and WAKELIN, R. W. (1953) Pyruvate oxidase system in brain tissue. *J. Physiol. (London), 119:* 421.

135. PETERS, R. A., WAKELIN, R. W., BUFFA, P., and THOMAS, L. C. (1953) Biochemistry of fluoroacetate poisoning. The isolation and some properties of the fluorotricarboxylic acid inhibitor of citrate metabolism. *Proc. Roy. Soc. (London), B, 140:* 497.

136. PETERS, R. A., WAKELIN, R. W., RIVETT, D. E. A., and THOMAS, L. C. (1953) Fluoroacetate poisoning: comparison of synthetic fluorocitric acid with enzymically synthesized fluorotricarboxylic acid. *Nature, 171:* 1111.

137. PETERS, R. A., and WILSON, T. H. (1952) A further study of the inhibition of aconitase by 'inhibitor fraction' isolated from tissues poisoned with fluoroacetate. *Biochim. Biophys. Acta, 8:* 348.

138. PETERS, R. A., and WILSON, T. H. (1952) A further study of the inhibition of aconitase by 'inhibitor fraction' isolated from tissues poisoned with fluoroacetate. *Biochim. Biophys. Acta, 9:* 310.

139. PHILLIPS, A. H., and LANGDON, R. G. (1955) Incorporation of

monofluoroacetic acid into the nonsaponifiable lipids of rat liver. *Arch. Biochem. Biophys., 58:* 247.

140. PRICE, C. C., and JACKSON, W. G. (1947) Some properties of methyl fluoroacetate and fluoroethanol. *J. Am. Chem. Soc., 69:* 1065.

141. QUIN, J. I., and CLARK, R. (1947) Studies on the action of potassium monofluoroacetate (CH_2FCOOK), [*Dichapetalum cymosum* (Hook) Engl.] toxin, on animals. *Onderstepoort J. Vet. Sci. Animal Ind., 22:* 77.

142. RAASCH, M. S. (1958) *α*, *α*-Difluoro-substituted acids of the tricarboxylic acid cycle, their salts, amides and esters. *U.S. Patent 2,824,888*, February 25, 1958.

143. REDEMANN, C. E., CHAIKIN, S. W., FEARING, R. B., ROTARIU, G. J., SAVIT, J., and VAN HOESEN, D. (1948) The vapor pressures of forty-one fluorine-containing organic compounds. *J. Am. Chem. Soc., 70:* 3604.

144. RIMINGTON, C. (1935) Chemical investigations of the 'Gifblaar' *Dichapetalum cymosum* (Hook) Engl. I. *Onderstepoort J. Vet. Sci. Animal Ind., 5:* 81.

145. RIVETT, D. E. A. (1953) The synthesis of monofluorocitric acid. *J. Chem. Soc., 1953:* 3710.

146. ROBERTS, J. D. (1959) *Nuclear magnetic resonance: applications to organic chemistry*. McGraw-Hill Book Co. Inc., New York.

147. SALLE, J. (1952) Sur les propriétés du bromure de fluoracétylcholine (F.ACH). *Arch. intern. pharmacodynamie, 91:* 339.

148. SAUNDERS, B. C. (1957) *Some aspects of the chemistry and toxic action of organic compounds containing phosphorus and fluorine*. Cambridge University Press, England.

149. SAUNDERS, B. C., and STACEY, G. J. (1948) Toxic fluorine compounds containing the C-F link. Part I. Methyl fluoroacetate and related compounds. *J. Chem. Soc., 1948:* 1773.

150. SAUNDERS, B. C., and STACEY, G. J. (1949) Toxic fluorine compounds containing the C-F link. Part IV. (a) 2-Fluoroethyl fluoroacetate and allied compounds. (b) 2:2′-Difluorodiethyl ethylene dithioglycol ether. *J. Chem. Soc., 1949:* 916.

151. SAUNDERS, B. C., STACEY, G. J., WILD, F., and WILDING, I. G. E. (1948) Esters containing phosphorus. Part V. Esters of substituted

phosphonic and phosphonous acids. *J. Chem. Soc., 1948:* 699.

152. SAUNDERS, B. C., STACEY, G. J., and WILDING, I. G. E. (1949) Toxic fluorine compounds containing the C-F link. Part II. 2-Fluoroethanol and its derivatives. *J. Chem. Soc., 1949:* 773.
153. SAUNDERS, B. C., and WILDING, I. G. E. (1949) Toxic fluorine compounds containing the C-F link. Part V. (a) Fluorine-containing ammonium salts. (b) Relationship between physiological action and chemical constitution. *J. Chem. Soc., 1949:* 1279.
154. SAUNDERS, B. C., and WORTHY, T. S. (1953) Toxic fluorine compounds containing the C-F link. Part IX. Preparation of carbon-labelled sodium fluoroacetate on a micro scale. *J. Chem. Soc., 1953:* 1929.
155. SCALES, J. W. (1945) '1080' rat poison very effective but dangerous later. *Miss. Agr. Expt. Sta., Farm Research, 8, No. 12:* 8.
156. SCHRADER, G. (1945) The development of new insecticides. Presented by MUMFORD, S. A., and PERREN, E. A. in *British Intelligence Objectives Sub-committee, Report No. 714.*
157. SCHRADER, G. *et al.* (1946) Developments in methods and materials for the control of plant pests and diseases in Germany. Reported by MARTIN, H., and SHAW, H. in *British Intelligence Objectives Sub-committee, Report No. 1095.*
158. SCHRADER, G. (1951) *Die Entwicklung neuer Insektizide auf Grundlage organischer Fluor- und Phosphor-Verbindungen*, Verlag Chemie, G.m.b.H., Weinheim, Germany.
159. SIDGWICK, N. V. (1933) *The covalent link in chemistry.* Cornell University Press, Ithaca, N.Y.
160. SPEYER, J. F., and DICKMAN, S. R. (1956) On the mechanism of action of aconitase. *J. Biol. Chem., 220:* 193.
161. STEYN, D. G. (1928) Gifblaar poisoning. A summary of our present knowledge in respect of poisoning by *Dichapetalum cymosum. 13th and 14th Reports of the Director of Veterinary Education and Research (Onderstepoort Laboratories)*, Part I, 187.
162. STEYN, D. G. (1934) Plant poisoning in stock and the development of tolerance. *Onderstepoort J. Vet. Sci. Animal Ind., 3:* 119.
163. STEYN, D. G. (1937) Poisonous plants. A serious problem in the Union of South Africa. *Farming in S. Africa, 12:* 80, 85.
164. STEYN, D. G. (1939) Plant poisoning in stock. II. Gifblaar poisoning. *Farming in S. Africa, 14:* 285.

165. STONER, H. B. (1956) The mechanism of toxic hepatic necrosis. *Brit. J. Exptl. Pathol., 37:* 176.

166. SWARTS, F. (1896) Sur l'acide fluoroacétique. *Bull. Acad. roy. Belg.,* [*3*], *31:* 675.

167. SWARTS, F. (1914) Sur l'alcool monofluoré et la fluoracétine éthylénique. *Bull. Acad. roy. Belg., Classe des Sciences, 1914:* 7.

168. THEILER, A. (1902) Unpublished results quoted by STEYN, D. G. (1928). *13th and 14th Reports of the Director of Veterinary Education and Research (Onderstepoort Laboratories)*, Part I, 187.

169. TIETZE, E., SCHEPSS, W., and HENTRICH, W. (1930) Verfahren zum Schützen von Wolle und dergleichen gegen Mottenfrass. *German Patent 504, 886*, August 9, 1930.

170. TOURTELLOTTE, W. W., and COON, J. M. (1951) Treatment of fluoroacetate poisoning in mice and dogs. *J. Pharmacol. Exptl. Therap., 101:* 82.

171. UMEDA, T. (1952) Sodium monofluoroacetate. *Japanese Patent 3874*, September 26, 1952.

172. WARD, J. C. (1956) How toxic are today's pesticides? *Pest Control, 24, No. 1:* 9, 12.

173. WARD, P. F. V., and PETERS, R. A. (1957) Experiments on the purification of fluorocitric acid. *Biochim. Biophys. Acta, 26:* 449.

174. WARD, P. F. V., and PETERS, R. A. (1958) Purification and transformation of synthetic fluorocitric acid. *Abstracts of communications presented at the IVth International Congress of Biochemistry, Vienna*, p. 3 (Section 1, Communication 19).

175. WATLAND, D. C., WANG, S. C., KALNITSKY, G., and HUMMEL, J. P. (1957) The inhibition of citrate formation from oxalacetate by ethyl esters of difluoroacetoacetate, fluorooxalacetate and fluoroacetate. *Arch. Biochem. Biophys., 67:* 138.

176. WATSON, H. B. (1941) *Modern theories of organic chemistry.* Oxford University Press, London.

177. WILLIAMS, A. T. (1948) Sodium fluoroacetate poisoning. *Hospital Corps Quarterly, 21:* 16.

178. WRIGHT, E. R. (1948) Personal communication to J. C. WARD, 1948.

3

ω-*Fluorocarboxylic acids and derivatives*

It was shown in the last chapter that fluoroacetic acid, FCH_2COOH and all compounds which can form this parent compound by some simple biochemical process such as hydrolysis or oxidation are highly toxic. While much of the work relating to these compounds was in progress, Kharasch and colleagues[27] in the United States prepared 3-fluoropropionic acid and 4-fluorobutyric acid*. The former was found to be non-toxic, whereas the latter was even more toxic than fluoroacetic acid.

FCH_2COOH	toxic
FCH_2CH_2COOH	non-toxic
$FCH_2CH_2CH_2COOH$	toxic

This highly significant observation gave rise to the suggestion that

---

* Conventionally, the position of substituents on the chain of aliphatic acids is indicated by numbering in systematic I.U.C. nomenclature, and by Greek lettering in trivial or common names; the numbering starts with the carbon of the COOH as 1, and the lettering with the carbon adjacent to the COOH as *α*. Thus,

$$Cl\overset{5}{C}H_2\overset{4}{C}H_2\overset{3}{C}H_2\overset{2}{C}H_2\overset{1}{C}OOH$$
$$\delta \quad \gamma \quad \beta \quad \alpha$$

5-chloropentanoic acid and *δ*-chloro-*n*-valeric acid are correct names for the formula shown. However, it is both common and more convenient to use numbering for *both* types of name, and this practice has been followed in this monograph. Examples of such names which are not strictly accurate include: 18-fluorostearic acid, $F(CH_2)_{17}COOH$ and methyl 2-fluoro-isobutyrate, $(CH_3)_2CFCOOCH_3$. The first five unbranched acids (formic to valeric) and a few very common higher acids (for example, stearic) are often referred to by their trivial names, whereas all others are usually named systematically.

a similar alternation in toxicity might be apparent on ascending the series still further. Consequently esters of many of the ω-fluorocarboxylic acids were prepared[8]. All the members examined provided striking confirmation of this suggestion. More recently the free acids have been obtained[42], the toxicological pattern of which is shown in Fig. 1 and Table II (pp. 5, 6). The results may be summarized briefly in the following important generalization: *in any ω-fluorocarboxylic acid,* $F(CH_2)_nCOOH$, *if the total number of carbon atoms in the chain is even, the compound is toxic, whereas if the total number is odd, the compound is non-toxic.* Simple derivatives of these acids have the same toxicological properties as the free acids themselves.

The pronounced difference in toxicity between 18-fluorostearic acid, $F(CH_2)_{17}COOH$ (toxic) and the isomeric 9-fluorostearic acid, $CH_3(CH_2)_8CHF(CH_2)_7COOH$ (non-toxic) emphasizes the high specificity of the ω-fluorine atom for pharmacological activity[46]. That fluorine *per se* is necessary for this activity is apparent from the fact that 10-fluorodecanoic acid, $F(CH_2)_9COOH$ is extremely toxic, whereas 10-chlorodecanoic acid, $Cl(CH_2)_9COOH$ is innocuous[42].

A full account of these and related compounds is given in the sections which follow. Members are listed in Table XI (p. 104).

OCCURRENCE

Dichapetalum toxicarium (Chailletia toxicaria, Don) is a glabrous (hairless) shrub occurring in Sierra Leone. Oliver has provided a complete description[35] in which it is recorded that the rather coriaceous (tough and leathery) leaves measure 2½ to 6 inches in length by 1¼ to 2¼ inches in breadth. The flowers are about 2 lines long, and the fruit, which is hard and woody, is 1 to 1½ inches long (see Fig. 6). The fruit is very poisonous, and has found use in a powdered form for killing rats. The common name of 'ratsbane' has therefore been coined for the shrub. In Sierra Leone it is also known as 'broke back' from its effect in producing paralysis of the lower limbs.

In two very colourful articles published in 1904[52, 53], Renner gives a vivid account of the poisonous effects of ratsbane, in which he observes that its use on humans by Country Doctors ('witch doctors') and others was far from uncommon. It has now been shown by Peters (p. 86) that the main toxic principle of ratsbane is a long-chain ω-fluorocarboxylic acid; hence the following description constitutes the sole authentic record of the general effect of these compounds on man. It is interesting to note incidentally that in Renner's reports, quoted below, reference is made to the use of ratsbane, and hence of an organic fluoroacid, as a rodenticide (anticipating Compound 1080 by over forty years) and as a chemical warfare agent (anticipating the proposed use of fluoroacetates for this purpose by about the same length of time). The description that follows is taken from the two above-mentioned reports of Renner.

'For some years now, death by poison has been the subject of talk in the Colony of Sierra Leone, and one could scarcely credit the statements so often made with respect to this subject. No one, it would appear, dies from natural causes. Poisoning in one form or another is put down as the cause of death not only amongst the poor but also amongst the rich, and yet no one would or could come forward to attest the fact that such a one has been poisoned by such a substance. Some of these deaths which appear to be mysterious have given cause for great alarm and anxiety and aroused a feeling of dread and bitterness against the Country Doctors. The Country Doctors, who have the true knowledge of the cause of the deaths, keep it a secret, attributing it to some mysterious influence of the devil or to witchcraft or to something occult, and thereby make much profit from their knowledge and wield great power over the masses.

'A peculiar kind of disease, of the origin of which no account could be given, is now and again met with among the people. The European Doctor, when called in, is puzzled, for he sees before him a young healthy man or woman, between the ages of 23 and 40 years, struck down with paralysis. He examines the case carefully, cross-examines scrupulously the patient and his friends, and fails to arrive at any conclusion from these as to the source of the trouble. He suspects poison, and is told the person has been poisoned. He endeavours to treat the symptoms, and either fails or the case is removed to a Country Doctor. The individual may live or die; this depending on the nature and the quantity of the poison administered.'

Such was the background for the discovery by Renner of the cause of this paralytic affliction. The case occurred in the person of a Mendi labourer aged about 24 years, by name Johnny, and the history was as follows: 'He was given to throw away some fish which had been poisoned by the ground fruit of ratsbane for the purpose of killing the house rats. He from greed ate the largest slice of the fish. In about half an hour afterwards he had vomiting followed by looseness of the bowels and general trembling. On his admission into hospital, he again vomited and his condition then as described by himself through an interpreter was as follows: 'He was feeling very weak and unable to walk, his legs were dead, and he was losing power over his arm and he was feeling very bad'. On examination in the ward it was found that the patient was suffering from complete paralysis of the lower extremities. The tendon reflexes were abolished. There was marked hyperaesthesia of the inner side of the thighs and legs. Firm pressure of the muscles of the calves gave unusually severe pain. The action of the bladder and rectum remained undisturbed. Although the pupils were normal, yet the patient's vision was not acute. There was some want of coordinating power in the muscles of the upper extremities. The power of deglutition was not affected. The patient's condition was stationary for about a fortnight, when he began to regain power over the lower limbs. He was able to stand with assistance, but was trembling all over, and on attempting to move he dragged his feet and presented the symptoms of one suffering from ataxia. He remained in hospital slowly improving and was discharged, about two months after admission, not completely well, there still remaining a certain amount of inability to use his limbs.'

Renner continues: 'The fruit of ratsbane is used largely for the destruction of rats and other animals; but beyond this it is also used extensively by the people in the Colony and the Hinterland to

References p. 108

poison one another. Conversing with some of the older men of the Timnes and Mendis, I found that with them it is very frequently used against their enemies to poison well-water or streams which supply hostile villages. Domestic animals poisoned by it are seen to rush about in great excitement as if in severe pain; they vomit and drag their hind limbs, which ultimately become paralyzed. They then lie down helplessly, breathing quickly, the forearm twitching and quivering; they ultimately die, apparently from paralysis of the respiratory muscles.'

Renner subsequently sent samples of the fruit of ratsbane from Sierra Leone to Power and Tutin in England. In the ensuing investigation, described in 1906[50], it was shown that the active principle was highly toxic. When 0.5 g of an aqueous extract was administered to a dog, a latent period of three hours was noticed; thereafter, characteristic delirium and epileptiform convulsions occurred, followed by death within 15 minutes. Post-mortem examination of the animal revealed a condition of cerebral congestion and thrombosis of the superior longitudinal sinus. An attempt at elucidating the structure of the poison was unsuccessful.

Some fifty years later, Peters reopened the subject and has reported some very interesting preliminary results[48]. It was found that symptoms of ratsbane poisoning were suggestive of poisoning by a C-F compound (high toxicity, convulsions, and accumulation of citric acid in the tissues). However there were several points of divergence from gifblaar (p. 13), the only other plant known to contain a C-F compound (fluoroacetic acid). Thus, the main toxic principle of gifblaar resided in the aqueous phase, whereas that of ratsbane was shown to be extracted by lipid solvents. Moreover, the distribution of accumulated citric acid after injection of the active principles was different for the two plants: gifblaar resulted in more citric acid in the kidney than in the heart, whereas ratsbane caused more in the heart than in the kidney; the former condition is typical of the simple fluoroacetates and the latter of long-chain ω-fluoro-acids (p. 89). In brief, the preliminary toxicological results indicated that the main active constituent of ratsbane might be a long-chain ω-fluorocarboxylic acid.

Separation of the fatty acid components by reversed phase partition chromatography showed that a major part of the active fraction behaved mainly as a C_{18} acid containing one double bond and one fluorine atom; from infrared analysis the latter was thought to be in the ω-position. Results so far available thus point to a fluoro-oleic acid. Although such a compound has not yet been synthesized in the laboratory, a few α-unsaturated ω-fluorocarboxylic acids have already been prepared (p. 99).

PREPARATIONS

Alkyl esters

The following methods are representative of those used for preparing the esters of the ω-fluorocarboxylic acids. An example of each reaction is shown, and references are given to indicate where experimental details may be found.

(a) The reaction of a bromo- or iodocarboxylate ester with pure, dry silver fluoride[8, 42]:

$$X(CH_2)_nCOOC_2H_5 + AgF \rightarrow F(CH_2)_nCOOC_2H_5 + AgX$$

(b) The reaction of a bromocarboxylate ester with potassium fluoride[40, 42]:

$$Br(CH_2)_nCOOC_2H_5 + KF \rightarrow F(CH_2)_nCOOC_2H_5 + KBr$$

(c) The esterification of the appropriate free acid[42]:

$$F(CH_2)_nCOOH + HOCH_3 \rightarrow F(CH_2)_nCOOCH_3 + H_2O$$

(d) The methanolysis of the appropriate nitrile[42, 46]:

$$F(CH_2)_nCN + H_2O + HOCH_3 \rightarrow F(CH_2)_nCOOCH_3 + NH_3$$

(e) The reaction of an ω-fluoroalkylmagnesium chloride with ethyl chloroformate[23, 42]:

$$F(CH_2)_nMgCl + ClCOOC_2H_5 \rightarrow F(CH_2)_nCOOC_2H_5 + MgCl_2$$

(f) The unsymmetrical anodic coupling of an ω-fluorocarboxylic acid with a half ester[42, 46, 47]:

$$F(CH_2)_nCOOH + HOOC(CH_2)_mCOOCH_3 \rightarrow F(CH_2)_{n+m}COOCH_3 + 2CO_2 + H_2$$

Acids

The free ω-fluorocarboxylic acids may be obtained by the following methods:

(a) The oxidation of the corresponding ω-fluoroalcohol by a variety of reagents, including potassium dichromate and sulphuric acid[42, 46], nitric acid [42], and chromium trioxide [42, 46]:

$$F(CH_2)_nCH_2OH + 2O \rightarrow F(CH_2)_nCOOH + H_2O$$

(b) The hydrolysis of an ester either by acid[42, 46] or by alkali[42]:

$$F(CH_2)_nCOOCH_3 + H_2O \rightarrow F(CH_2)_nCOOH + CH_3OH$$

(c) The hydrolysis of the appropriate nitrile[40, 42]:

$$F(CH_2)_nCN + 2H_2O \rightarrow F(CH_2)_nCOOH + NH_3$$

(d) The reaction of an ω-fluoroalkylmagnesium chloride or bromide with carbon dioxide[39, 42]:

$$F(CH_2)_nMgX + CO_2 + H_2O \rightarrow F(CH_2)_nCOOH + Mg(OH)X$$

(e) Special methods, for example[32, 34]:

$$(CH_2)_5\begin{matrix} \diagup CO \\ | \\ \diagdown NH \end{matrix} \xrightarrow{HNO_2} (CH_2)_5\begin{matrix} \diagup CO \\ | \\ \diagdown N\text{-}NO \end{matrix} + HF \rightarrow F(CH_2)_5COOH + N_2$$

$$CH_2{=}CH_2 + CO_2 + HF \xrightarrow[FeF_3]{BF_3} FCH_2CH_2COOH$$

The esters are pleasant-smelling liquids and the acids are colourless liquids or solids. Members are listed in Table XI (p. 104). The fluorine atom is very inert, with the result that the chemical properties of these esters and acids are similar to those of the corresponding unfluorinated compounds.

TOXICOLOGY

The toxicological pattern induced by the long-chain ω-fluorocarboxylic acids bears a strong resemblance to that of fluoroacetic acid itself *(q.v.)*. Points in common include high toxicity, charac-

Fig. 6. Dichapetalum toxicarium (Chailletia toxicaria, Don.*).*
Photographs of dry dupe (the nut-like fruit containing the seeds) and of the separated seeds.

(Reproduced with permission, R. A. Peters, *Endeavour, 13* (1954) 147.)

teristic symptoms, and accumulation of citric acid in the tissues; detailed clinical data relating to fluoroacetate poisoning in a variety of animals are given on p. 29, and in man on p. 47. The only authentic information on the direct effects of a long-chain ω-fluoroacid on man is that provided by Renner; this is quoted in full on p. 85.

This account would be less than complete, however, were no mention made of some differences in response between fluoroacetic acid on the one hand and the higher ω-fluoroacids on the other. Not only are the higher members more toxic (reasons for this are discussed on p. 94) but they act more rapidly and cause a more pronounced increase in the rate of the heart beat. Moreover, there is a definite difference in the distribution of accumulated citric acid, in that the amounts in the heart and kidney are usually reversed; this is illustrated in Table VII:

TABLE VII*

CITRATE ACCUMULATION AFTER POISONING WITH FLUOROACIDS

	μg/g wet tissue	
	Heart	*Kidney*
Fluoroacetate (5–10 mg/kg)	677	1000
Fluorobutyrate (3 mg/kg)	1045	386
Fluorodecanoate (10 mg/kg)	1070	240

Finally, the effect of the long-chain ω-fluoro-acids on the accumulation of citric acid *in vitro* is very much more pronounced; for example, as little as 0.0006 mg of 10-fluorodecanoic acid produced a rise in citric acid content in kidney particles, whereas at least 0.025 mg of fluoroacetate would be required to show a similar response; on a molar basis, this difference is about 100-fold[48]. A similar effect is apparent *in vivo* (see Table VIII, p. 94).

Curious biological anomalies continue to appear in the literature.

* Compiled by R. A. Peters[48].

For example, Chenoweth[10] has observed that, whereas 4-fluorocrotonate resembles fluoroacetate, 4-fluorobutyrate is quite dissimilar to fluoroacetate in its pharmacodynamic properties; again, 4-fluorobutyric acid has been reported to inhibit the growth of *Escherichia coli* in a manner both qualitatively and quantitatively identical to that observed with fluoroacetic acid, while 6-fluorohexanoic acid was completely inert[30].

β-OXIDATION

In 1904 Knoop proposed[28] the theory of β-oxidation to explain the metabolic degradation of fatty acids. Essentially this involves oxidation at the β-carbon atom followed by hydrolytic chain cleavage. For example, hexanoic acid would be catabolized by the following stages:

$$
\begin{aligned}
CH_3CH_2CH_2\overset{\beta}{C}H_2\overset{\alpha}{C}H_2COOH &\rightarrow CH_3CH_2CH_2CO \vdots CH_2COOH \\
&\rightarrow CH_3\overset{\beta}{C}H_2\overset{\alpha}{C}H_2COOH + CH_3COOH \\
&\rightarrow CH_3CO \vdots CH_2COOH \\
&\rightarrow CH_3COOH + CH_3COOH
\end{aligned}
$$

This scheme is of course an oversimplification, since it does not account for the intimate mechanism of the oxidative and hydrolytic steps; nor does it indicate the source of biochemical activation, now known to be effected by the initial formation of the thiol-ester of coenzyme A. Nevertheless the overall process has not seriously been challenged during the half-century since it was originally proposed. In short, it is now accepted that fatty acids are degraded by the successive or simultaneous removal of two-carbon fragments.

If this process of β-oxidation be applied to the ω-fluorocarboxylic acids, a ready explanation is apparent for the pronounced alternation in toxicity described above (p. 82). It can be seen that all the toxic members (*i.e.* those containing an even number of carbon atoms) can form the very toxic fluoroacetic acid, whereas the non-toxic members (*i.e.* those containing an odd number of carbon atoms) can be degraded only so far as the non-toxic 3-fluoropropionic acid (or its non-toxic catabolites). The mode of action of fluoroacetic acid, once formed, is discussed in detail in Chapteı 2. Strictly speaking, it is not necessary to postulate the actual intermediate formation of fluoroacetic acid itself, since 'activated fluoroacetate' (fluoroacetyl-coenzyme A) may be formed directly. The two examples that follow illustrate this general explanation:

$$FCH_2CH_2CH_2CH_2COOH \text{ (non-toxic)} \rightarrow FCH_2CH_2COCH_2COOH \rightarrow FCH_2CH_2COOH \text{ (non-toxic)}$$

$$FCH_2CH_2CH_2CH_2CH_2COOH \text{ (toxic)} \rightarrow FCH_2CH_2CH_2COCH_2COOH \rightarrow FCH_2CH_2CH_2COOH \rightarrow FCH_2COCH_2COOH \rightarrow FCH_2COOH \text{ (toxic)}$$

The corollary to the explanation is that the toxicity pattern provides novel and independent verification of the β-oxidation theory. This aspect of the subject is developed more thoroughly in Chapter 5 (p. 182).

Evidence for the β-oxidation theory as outlined above has been obtained by two means: (a) examination of the toxicological properties of fluoroacids in which β-oxidation is inhibited by structural alterations; and (b) biochemical studies using the free fluoroacids. These two independent lines of investigation are described below.

(a) Structural alterations

This method consisted in synthesizing ω-fluoro compounds which contained the 'skeleton' of the toxic members, but in which β-oxidation was inhibited by appropriate structural means[45]. The simplest example of a compound in which the β-position is effectively 'blocked' is ethyl 4-fluoro-3,3-dimethylbutyrate:

$$FCH_2C(CH_3)_2CH_2COOEt$$

This was synthesized[45] for comparison with the highly toxic ethyl 4-fluorobutyrate:

$$FCH_2CH_2CH_2COOEt$$

It was argued that 'blockage' of the β-carbon atom should render β-oxidation impossible, and hence should result in a non-toxic compound; in other words, if the 'blocked' compound *were* toxic, then clearly β-oxidation was *not* the true explanation. However on injection into mice it was found to be entirely devoid of toxic properties, thus providing support for the β-oxidation theory outlined above.

An alternative means of inhibition, involving the incorporation of the α- and β-carbon atoms into a ring system, led to the same conclusion. For comparison with the above-mentioned ethyl 4-fluorobutyrate, four compounds exemplified by the following formulae were prepared (p. 203) and examined[45]:

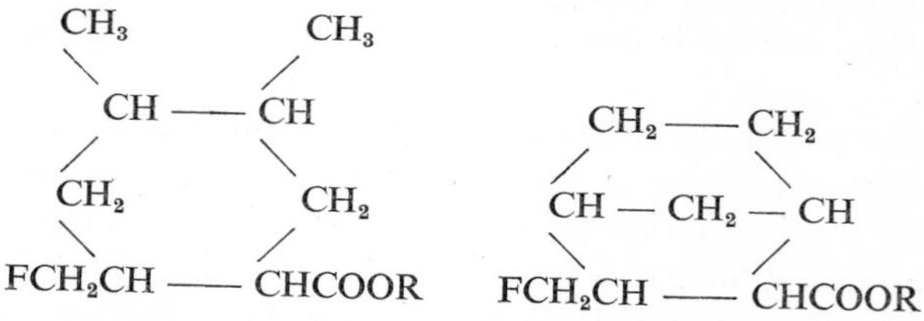

$$\begin{array}{cccc} CH_3 & & & CH_3 \\ & \diagdown & \diagup & \\ & CH & — CH & \\ & \diagup & \diagdown & \\ CH_2 & & & CH_2 \\ & \diagdown & \diagup & \\ & FCH_2CH & — CHCOOR & \end{array} \qquad \begin{array}{ccccc} & CH_2 & — & CH_2 & \\ \diagup & & & & \diagdown \\ CH & — & CH_2 & — & CH \\ \diagdown & & & & \diagup \\ & FCH_2CH & — & CHCOOR & \end{array}$$

It was considered unlikely that the animal body could degrade ring compounds of this type by a process of β-oxidation. The fact that all such compounds were non-toxic therefore provided additional confirmation, by the type of argument given above, of the β-oxidation theory.

Other variations of this method were examined, involving a variety of aliphatic ω-fluoroacids containing an etheric oxygen atom as a chain member[9]. The toxicity results obtained with these compounds are rather confused; simple rupture of the ether link[41] may be just as significant as the arguments revolving around β-oxidation, so the discussion of these compounds is more appropriately included along with that of simple fluoroethers (Chapter 4, p. 138).

In regard to all the compounds involving structural alterations in the ω-fluorocarboxylic acid series, Chenoweth[10] very properly voices a warning against placing too much reliance on results obtained using such biologically abnormal compounds; he points out that their structural deformity as such may hinder the compounds from entering a reactive centre, and for that reason are pharmacologically inert. Consequently the biochemical investigations described below provide more reliable and convincing evidence of the β-oxidation theory.

(b) Biochemical studies

As described in Chapter 2, one of the consequences of fluoroacetate poisoning is the accumulation of citric acid in various tissues. This accumulation has been explained as resulting from the biochemical conversion of fluoroacetate to fluorocitric acid, which in turn is a strong competitive inhibitor of the enzyme (aconitase) responsible for the normal utilization of citric acid *in vivo* (p. 42). Since there is no evidence to suggest that the long-chain acids have any effect *per se* on aconitase, any fluoro compound which results in marked citric acid accumulation may therefore be considered to have been metabolized to fluoroacetic acid (or, more precisely, to fluoroacetyl-coenzyme A). This then affords an excellent independent means of checking the β-oxidation theory, since, if valid, the toxic fluoroacids should form fluoroacetyl-coenzyme A and hence give rise to a pronounced accumulation of citric acid, while the non-toxic fluoroacids should not. This general technique has been used by Kandel and Chenoweth[25] and later by Parker and Walker[36, 57] for the examination of a wide variety of ω-fluoro compounds, including the members of the ω-fluorocarboxylic acid series now under discussion. The results shown in Table VIII are those obtained by Parker and Walker[36].

The toxicity figures listed in the first column confirm that the even fluoroacids are very much more toxic than the odd fluoroacids. But the most striking feature is the fact that the even fluoroacids caused a very large accumulation of citrate, whereas the odd fluoroacids caused little or none; indeed, the citric acid levels after dosage with

TABLE VIII

THE TOXICITY AND CITRIC ACID ACCUMULATION CAUSED BY REPRESENTATIVE ω-FLUOROCARBOXYLIC ACIDS ADMINISTERED TO MICE

	LD_{50} (95% confidence limits) μmole/kg	Dose for citrate study μmole/kg	Citrate concentration in kidney μg/g tissue
Sodium fluoroacetate	129 (95.5–174)	164	766
3-Fluoropropionic acid	553 (501–620)	2290	38
6-Fluorohexanoic acid	11.8 (3.10–44.7)	41	1414
9-Fluorononanoic acid	681 (552–827)	2150	112

the odd fluoroacids showed the same variation as those in kidneys from untreated animals. It is also noteworthy to compare the massive dosage of the odd fluoroacids where no citrate accumulation occurred with the small dosage of the even fluoroacids where there was marked citrate accumulation. These observations are taken as further evidence in support of the hypothesis that the toxic ω-fluorocarboxylic acids are metabolized to fluoroacetate, while the non-toxic acids are not. In short, the results are in excellent accord with the β-oxidation theory.

Further evidence for β-oxidation is provided by the toxicity figures of some of the derivatives supposedly formed in the process; these include modified ω-fluorocarboxylates of the following types: $F(CH_2)_nCH{=}CHCOOR$, $F(CH_2)_nCH(OH)CH_2COOR$ and $F(CH_2)_nCOCH_2COOR$. The argument is developed on p. 103.

There is one feature connected with these toxicities of the ω-fluorocarboxylic acid series which is not readily explained by the β-oxidation theory. One would expect, as the series is ascended, that less and less fluoroacetate would be produced from any given weight of an ω-fluorocarboxylic acid; that is, that the magnitude of the toxicity should decrease. Strangely, the reverse is true[8, 42]. It may be that the longer-chain acids, due to their greater lipid solubility, can penetrate the cells more readily, there giving rise to a higher concentration of the toxic principle (fluoroacetyl-coenzyme A)[8]; or that, in the process of β-oxidation, a higher concentration of 'active fluoroacetate' (fluoroacetyl-coenzyme A) may be available from the long-chain members than from fluoroacetic acid itself (due possibly to an adverse intramolecular electronic influence of the α-fluorine atom in forming fluoroacetyl-coenzyme A); or that both these effects occur directly in some area where they could do most harm. Whatever the cause, it should be noted that this anomalous situation, while not invalidating the β-oxidation theory, leads to the suggestion that other factors may also be involved; Chenoweth[10], for example, considers that a long-chain ω-fluorocarboxylic acid may interfere significantly with the metabolism of the corresponding non-fluorinated fatty acid in the cell. From the evidence at present available, the author favours the β-oxidation mechanism on which side effects, as yet unproved, may be superimposed.

PHARMACOLOGICAL ASPECTS

Antagonists and prophylactics

The most successful antidote and prophylactic for *fluoroacetate* poisoning involves massive intramuscular injections of monoacetin (glycerol monoacetate, glyceryl monoacetate)[12]; oral administration is effective to a lesser degree[11]. Acetamide has also been found to reduce mortality[18, 19]. The protective action of both compounds is considered to be due to the liberation of 'active acetate' (acetyl-coenzyme A) for competition with the toxic 'active fluoroacetate' (fluoroacetyl-coenzyme A). It is not surprising therefore that the most successful antagonist of 4-fluorobutyric acid is monobutyrin (glyceryl monobutyrate)[22, 25], presumably for the same reason. It is of considerable interest that, while monoacetin affords specific protection against fluoroacetate and monobutyrin against 4-fluorobutyrate, no cross protection has been observed; that is, monoacetin does not antagonize 4-fluorobutyrate and monobutyrin does not antagonize fluoroacetate. That monoacetin does not protect against fluorobutyrate may be explained[48] in terms of the formation of fluoroacetyl-coenzyme A from the latter *without* the actual intermediate formation of fluoroacetate itself; thus the stage at which the antagonistic action of monoacetin takes place is in effect

References p. 108

by-passed. No mention appears in the literature of work with glyceryl monohexanoate, which would be interesting both in its own right and for resolving the difficulties outlined in the next paragraph. The results so far available therefore indicate that the monoacetin therapy may not be the most effective treatment against poisoning by the long-chain ω-fluorocarboxylates.

Experiments using free fatty acids as potential antagonists present a more confused picture. In an early study, Kalnitsky and Barron[24] using kidney suspensions observed that 4-fluorobutyrate was more powerful than fluoroacetate in inhibiting butyrate oxidation, while fluoroacetate was more powerful than fluorobutyrate in inhibiting acetate oxidation. In the light of more recent work, these results could be interpreted as indicating a specific antagonism of 4-fluorobutyrate by butyrate and of fluoroacetate by acetate; this is therefore in line with the work with monoacetin and monobutyrin described above. More recently, however, Hendershot and Chenoweth[21], using the technique involving intestinal mobility[14, 15], have demonstrated that, although fluoroacetate is best antagonized by acetate as expected, 4-fluorobutyrate *and* 6-fluorohexanoate are both best antogonized by hexanoate. Arguing by analogy with the accepted mechanism of fatty acid metabolism[3, 4, 31], Hendershot and Chenoweth have suggested that acetate on the one hand and the higher fatty acids on the other may exert their specific antagonistic effects at different points in the overall metabolic pathway; thus, acetate may antagonize fluoroacetate in its conversion into fluoroacetyl-coenzyme A (that is, just prior to its entry into the tricarboxylic acid cycle), whereas butyrate and hexanoate may antagonize the higher fluoroacids during their β-oxidation (which has been discussed above, p. 90). This explanation can be extended to account for the lack of cross protection with the glyceryl esters, mentioned above.

Tolerance

The ability of small doses of fluoroacetate to protect against subsequent large doses is discussed in Chapter 2 (p. 36); it was shown that the effect can be produced only within certain limits and in certain species of animals. Kandel and Chenoweth[26] have extended this information to include 4-fluorobutyric acid, the only long-chain ω-fluorocarboxylic acid so far examined. The unexpected results obtained were: (a) that small doses of fluoroacetate increased the resistance in rats to challenging doses of both fluoroacetate and 4-fluorobutyrate; but (b) that small doses of 4-fluorobutyrate evoked *no* resistance in rats to challenging doses of fluoroacetate and of 4-fluorobutyrate.

Cumulation

The cumulative effect of fluoroacetates has been described in Chapter 2 (p. 36). The only information available on the long-chain ω-fluorocarboxylic acids is a passing reference to the cumulative effect of the toxic principle of ratsbane (p. 83) and hence of a toxic fluoroacid. In 1906, Power and Tutin pointed out that 'the poison which causes convulsions' is very slowly excreted by the dog, with the result that death could be induced by the administration of a series of individually innocuous doses[50].

MISCELLANEOUS DERIVATIVES OF ω-FLUOROCARBOXYLIC ACIDS

2-Fluoroethyl esters

The simple alkyl esters have been described above (p. 87). In an attempt to enhance their toxicity, certain 2-fluoroethyl esters of the toxic members were prepared[8] by the reactions:

$$X(CH_2)_nCOOH \rightarrow X(CH_2)_nCOOCH_2CH_2F \xrightarrow{AgF} F(CH_2)_nCOOCH_2CH_2F$$
$$(X = Br \text{ or } I)$$

These esters were expected on hydrolysis *in vivo* to result in the formation of two very potent entities: 2-fluoroethanol and the parent toxic ω-fluorocarboxylic acid. The results shown in Table IX indicate that some enhancement of toxicity occurs with the lower members. It should be mentioned however that all the four simple *alkyl* esters listed have been prepared more recently by different methods[42] and that all were found to be much more toxic than was originally thought. The newer figures are shown in brackets in Table IX. It is of course possible that this discrepancy is due to such causes as strain differences in the mice, but caution must be exercised in drawing too definite conclusions from the figures as they stand.

TABLE IX

TOXICITIES OF ETHYL AND 2-FLUOROETHYL ESTERS

	LD_{50} for mice (mg/kg) (intraperitoneal, propylene glycol as solvent)	
	$R = C_2H_5$-	$R = FCH_2CH_2$-
FCH_2COOR	15* (7.5*)	8.5
$F(CH_2)_5COOR$	4 (1.61*)	2.5
$F(CH_2)_7COOR$	9 (1.75)	7
$F(CH_2)_9COOR$	10 (1.65)	10

* Methyl ester

It can be seen from the original figures listed in the Table that the 2-fluoroethyl moiety apparently increases the toxicity of the *lower* ω-fluorocarboxylates, but that the effect gets less on ascending the homologous series until it is absent in the 10-fluorodecanoate esters. Two points have been tentatively put forward[8] to account for this observation. (a) As the series is ascended, the amount of 2-fluoroethanol available from a given weight of ester must decrease; it is suggested that a point will eventually be reached when the amount of liberated 2-fluoroethanol will be too small to make any apparent difference in the toxicity. (b) It is possible that the 2-fluoroethyl ω-fluorocarboxylates may exert some subsidiary action *per se*, independently of any subsequent degradation (this action would of necessity be associated with the two terminal fluorine atoms); it is suggested that there may be an optimum stereochemical distance separating the fluorine atoms for maximum activity, and that, if so, this distance must approximate that of the *shorter* chains.

Acid halides, anhydrides, amides, and anilides

It was anticipated[38] that the toxicity of the acid halides, anhydrides, amides and anilides would be of the same order as that of the parent acids, due to hydrolysis *in vivo*. The results obtained from the very limited number tested confirm this: for example, the LD_{50} by intraperitoneal injection into mice of 6-fluorohexanoamide, $F(CH_2)_5CONH_2$ was 0.7 mg/kg (*cf.* 6-fluorohexanoic acid, 1.35 mg/kg) and of 3-fluoropropionic anhydride, $F(CH_2)_2COOCO(CH_2)_2F$ was 137 mg/kg (*cf.* 3-fluoropropionic acid, 60 mg/kg). These derivatives were all prepared from the free acids by standard means[38].

Unsaturated acids and esters

Only three unsaturated acids or esters have so far been prepared (Table XI, p. 106). Esters of 4-fluorocrotonic acid, $FCH_2CH{=}CHCOOR$ have been obtained by the following methods:

(a) $$FCH_2\overset{O}{CH - CH_2} \xrightarrow{HCN} FCH_2CH(OH)CH_2CN \rightarrow FCH_2CH(OH)CH_2COOCH_3 \rightarrow FCH_2CHClCH_2COOCH_3 \rightarrow FCH_2CH{=}CHCOOCH_3 \quad (27)$$

(b) $$BrCH_2CH{=}CHCOOCH_3 + AgF \rightarrow FCH_2CH{=}CHCOOCH_3 + AgBr \quad (45)$$

(c) $$CH_3SO_2OCH_2CH{=}CHCOOC_2H_5 + KF \rightarrow FCH_2CH{=}CHCOOC_2H_5 + CH_3SO_2OK \quad (43)$$

Two higher members, 12-fluorododec-2-enoic acid, $F(CH_2)_9CH{=}CHCOOH$ and 13-fluorotridec-2-enoic acid, $F(CH_2)_{10}CH{=}CHCOOH$ were prepared[42, 61] by the Knoevenagel reaction:

(d) $$F(CH_2)_nCHO + H_2C(COOH)_2 \rightarrow F(CH_2)_nCH{=}CHCOOH + CO_2$$

It has been reported that the crotonate esters are very toxic[8, 27, 42], and that they are much more rapid in their lethal action than fluoroacetate at equivalent concentrations[10]. All three members

References p. 108

Univ. of Arizona Library

have toxicities similar to those of the corresponding saturated ω-fluorocarboxylates; for example[42], ethyl 4-fluorocrotonate has an LD_{50} in mice of 1.25 mg/kg (*cf.* methyl 4-fluorobutyrate, 0.7 mg/kg) and 12-fluorododec-2-enoic acid has an LD_{50} of 1.55 mg/kg (*cf.* 12-fluorododecanoic acid, 1.25 mg/kg). These facts are consistent with modern concepts of fatty acid metabolism, as outlined below (p. 103).

Hydroxyacids and esters

Ethyl ω-fluorolactate, $FCH_2CH(OH)COOC_2H_5$ and methyl 4-fluoro-3-hydroxybutyrate, $FCH_2CH(OH)CH_2COOCH_3$ were both obtained from epifluorohydrin[44] (Table XI, p. 106). The former was prepared by oxidation with nitric acid followed by esterification[44]; and the latter by treatment with hydrogen cyanide forming 4-fluoro-3-hydroxybutyronitrile, followed by methanolysis[27]:

$$FCH_2\overset{\quad O \quad}{CH - CH_2} \rightarrow FCH_2CH(OH)COOH \rightarrow FCH_2CH(OH)COOC_2H_5$$

$$FCH_2\overset{\quad O \quad}{CH - CH_2} \rightarrow FCH_2CH(OH)CH_2CN \rightarrow FCH_2CH(OH)CH_2COOCH_3$$

Two hydroxyacids containing non-terminal fluorine atoms (Table XI, p. 106) have been prepared, both from diethyl fluoro-oxalacetate (p. 101). Monofluorocitric acid, $HOOCCHFC(OH)(COOH)CH_2COOH$ was obtained by the Reformatski reaction[54, 58], and dimethyl fluoromalate, $CH_3OOCCHFCH(OH)COOCH_3$, by reduction with potassium borohydride [55, 56]:

$$EtOOCCHFCOCOOEt \xrightarrow[BrCH_2COOEt]{Zn} EtOOCCHFC(OH)(COOEt)CH_2COOEt$$

$$\xrightarrow{hydr.} HOOCCHFC(OH)(COOH)CH_2COOH$$

$$EtOOCCHFCOCOOEt \xrightarrow[CH_3OH]{KBH_4} CH_3OOCCHFCH(OH)COOCH_3$$

Ethyl ω-fluorolactate was found to be non-toxic, as was expected from the low toxicity of the closely related fluoropyruvic acid, $FCH_2COCOOH$[6, 29] and of the parent 3-fluoropropionic acid, FCH_2CH_2COOH. Methyl 4-fluoro-3-hydroxybutyrate on the other hand was very toxic[10, 27]. These results are consistent with the generally accepted mechanism of β-oxidation, as outlined below (p. 103). Fluorocitric acid has been discussed in Chapter 2 (p. 42); it has now been found that the synthetic material contains varying mixtures of up to twelve substances[59, 60]. Dimethyl fluoromalate was found to be non-toxic[48]; the disodium salt has been reported to inhibit malic and malic dehydrogenase enzymes[16].

Ketoacids and esters

Six ketoacids or esters have been prepared (Table XI, p. 106). Fluoropyruvic acid, $FCH_2COCOOH$[6, 29, 33] was obtained from ethyl fluoroacetate by conversion to diethyl fluoro-oxalacetate[6, 54] followed by hydrolytic decarboxylation:

$$EtOOCCH_2F + EtOOCCOOEt \rightarrow EtOOCCHFCOCOOEt \rightarrow FCH_2COCOOH$$

The other four members were prepared[13] by the elegant method of Bowman and Fordham[7]. This involved the reaction of ethyl tetrahydropyranyl sodiomalonate with the appropriate acid chloride, with subsequent removal of the tetrahydropyranyl group and decarboxylation:

$$\begin{aligned} F(CH_2)_nCOCl + NaCH(COOEt)COOC_5H_9O \\ \rightarrow F(CH_2)_nCOCH(COOEt)COOC_5H_9O \\ \rightarrow F(CH_2)_nCOCH(COOEt)COOH \\ \rightarrow F(CH_2)_nCOCH_2COOH \ (n = 1, 5, 6 \text{ and } 9) \end{aligned}$$

The toxicity of the members is shown in Table X. For comparison, the figures for the parent ω-fluorocarboxylic acid or ester are included.

TABLE X

TOXICITIES OF SOME FLUOROKETOCARBOXYLATES

LD_{50} for mice (intraperitoneal, propylene glycol as solvent) mg/kg			
$FCH_2COCOOH$	80	FCH_2CH_2COOH	60
$C_2H_5OOCCHFCOCOOC_2H_5$	>400	$NaOOCCHFCH_2COONa$	non-toxic
$FCH_2COCH_2COOC_2H_5$	2.5	$F(CH_2)_3COOCH_3$	0.7
$F(CH_2)_5COCH_2COOC_2H_5$	1.3	$F(CH_2)_7COOC_2H_5$	1.75
$F(CH_2)_6COCH_2COOC_2H_5$	67	$F(CH_2)_8COOC_2H_5$	70
$F(CH_2)_9COCH_2COOC_2H_5$	1.95	$F(CH_2)_{11}COOH$	1.25

It is at once apparent that the toxicities of the ω-fluoroketocarboxylates are closely paralleled by those of the corresponding ω-fluorocarboxylates, thus establishing the biochemical similarity of the two classes. This is consistent with the theory that β-keto-acids are intermediary metabolites of fatty acids, as outlined below.

Mention must be made of fluoropyruvic acid and its probable biochemical action. Since pyruvic acid is the main precursor of the activated C_2 fragment necessary for incorporation into the tricarboxylic acid cycle, particular attention has been directed to the mode of action of its fluoro analogue. The relatively low toxicity of fluoropyruvic acid and its failure to produce convulsions effectively exclude the obvious mechanism involving conversion to fluoroacetyl-coenzyme A with subsequent formation of fluorocitric acid; this process is further excluded by the lack of any appreciable accumulation of citric acid *in vivo* or *in vitro*[17]. Avi-Dor and Mager[1, 2] have now suggested convincingly that fluoropyruvic acid operates by a biochemical reaction akin to that of bromo- and iodoacetic acids with thiol compounds (p. 25):

$$RSH + FCH_2COCOOH \rightarrow RSCH_2COCOOH + HF$$

Thus the vicinal carbonyl group of the fluoropyruvic acid apparently decreases the stability of the normally inert C-F bond to such an extent that the lability of the fluorine atom approximates that of the halogen atoms in bromo- and iodoacetic acids. Evidence for the above mechanism resides in the formation of the thioether link with simultaneous disappearance of the thiol group. Unequivocal proof of the liberation of free fluoride in the reaction has recently been supplied by Peters and Hall[49].

Diethyl fluoro-oxalacetate, EtOOCCHFCOCOOEt, a ketoester containing a non-terminal fluorine atom, is non-toxic [5, 6, 37] and produces only a slight citrate accumulation[17]. Hence it apparently does not enter the tricarboxylic acid cycle to form fluorocitric acid. The corresponding sodium monofluorosuccinate, $NaOOCCHFCH_2$-COONa has been reported also to be non-toxic[20]; however, the identity of this material and of its esters is in some doubt[48] (see also Chapter 2, ref. 12).

DISCUSSION

The derivatives (unsaturated, hydroxy- and keto-) described above and listed in Table XI (p. 106) are all relevant to the study of the more detailed mechanism of β-oxidation. It is probable that this occurs by a route such as:

$$RCH_2CH_2COOH \xrightarrow{-2H} RCH{=}CHCOOH \xrightarrow{H_2O} RCH(OH)CH_2COOH$$
$$\xrightarrow{-2H} RCOCH_2COOH \xrightarrow{H_2O} RCOOH + CH_3COOH$$

activation of the original acid being effected by the initial formation of the thiol-ester of coenzyme A; all such changes are reversible, but for convenience are shown as unidirectional. The fact that these various derivatives of the ω-fluorocarboxylates have approximately the same toxicity as the parent compounds provides indirect but substantial evidence that these derivatives are formed at least transitorily in the metabolism of the ω-fluorocarboxylates themselves. If on the other hand there had been pronounced discrepancies in the toxicity patterns, then the scheme as outlined above would require modification. By extending the conclusions to unfluorinated members, the results described in this chapter afford independent evidence for the mode of breakdown of fatty acids *in vivo*.

References p. 108

TABLE XI*

Compound	*Formula*
(a) *ω-Fluorocarboxylic acids*	
Fluoroacetic acid	FCH_2COOH
3-Fluoropropionic acid	$F(CH_2)_2COOH$
4-Fluorobutyric acid	$F(CH_2)_3COOH$
5-Fluorovaleric acid	$F(CH_2)_4COOH$
6-Fluorohexanoic acid	$F(CH_2)_5COOH$
7-Fluoroheptanoic acid	$F(CH_2)_6COOH$
8-Fluoro-octanoic acid	$F(CH_2)_7COOH$
9-Fluorononanoic acid	$F(CH_2)_8COOH$
10-Fluorodecanoic acid	$F(CH_2)_9COOH$
11-Fluoroundecanoic acid	$F(CH_2)_{10}COOH$
12-Fluorododecanoic acid	$F(CH_2)_{11}COOH$
18-Fluorostearic acid	$F(CH_2)_{17}COOH$
(b) *ω-Fluorocarboxylate alkyl esters*	
Methyl 3-fluoropropionate	$F(CH_2)_2COOCH_3$
Methyl 4-fluorobutyrate	$F(CH_2)_3COOCH_3$
Methyl 5-fluorovalerate	$F(CH_2)_4COOCH_3$
Ethyl 5-fluorovalerate	$F(CH_2)_4COOC_2H_5$
Methyl 6-fluorohexanoate	$F(CH_2)_5COOCH_3$
Ethyl 6-fluorohexanoate	$F(CH_2)_5COOC_2H_5$
Methyl 7-fluoroheptanoate	$F(CH_2)_6COOCH_3$
Ethyl 7-fluoroheptanoate	$F(CH_2)_6COOC_2H_5$
Ethyl 8-fluoro-octanoate	$F(CH_2)_7COOC_2H_5$
Ethyl 9-fluorononanoate	$F(CH_2)_8COOC_2H_5$
Ethyl 10-fluorodecanoate	$F(CH_2)_9COOC_2H_5$
Ethyl 11-fluoroundecanoate	$F(CH_2)_{10}COOC_2H_5$
Ethyl 16-fluorohexadecanoate	$F(CH_2)_{15}COOC_2H_5$
Methyl 18-fluorostearate	$F(CH_2)_{17}COOCH_3$
(c) *ω-Fluorocarboxylate 2-fluoroethyl esters*	
2-Fluoroethyl 6-fluorohexanoate	$F(CH_2)_5COOCH_2CH_2F$
2-Fluoroethyl 8-fluoro-octanoate	$F(CH_2)_7COOCH_2CH_2F$
2-Fluoroethyl 10-fluorodecanoate	$F(CH_2)_9COOCH_2CH_2F$

ω-FLUOROCARBOXYLIC ACIDS** AND DERIVATIVES

LD_{50} *(mice, I.P.) mg/kg*	*Conclusion*	*Boiling point*	*References*
6.6	toxic	167–168°	42
60	indefinite	83–84°/14 mm	42
0.65	very toxic	78–79°/6 mm	42
> 100	non-toxic	90°/4 mm	42
1.35	very toxic	114°/6 mm	42
40	indefinite	133–134°/10 mm	42
0.64	very toxic	145–148°/10 mm	42
> 100	non-toxic	100–101°/0.15 mm	42
1.5	very toxic	m.p. 49–49.5°	42
57.5	indefinite	m.p. 36–36.5°	42
1.25	very toxic	m.p. 59.5–61°	42
5.7	toxic	m.p. 68.5–69°	42
—	non-toxic	118–120°	42
0.7	very toxic	78.5°/100 mm	42, 51
> 100	non-toxic	72–74°/25 mm	42
> 160	non-toxic	56–60°/16 mm	8
1.6	very toxic	70–71°/9 mm	42
4	toxic	82–84°/14 mm	8
> 100	non-toxic	92–93°/13 mm	42
—	non-toxic	97–97.5°/11 mm	42
1.75	very toxic	106.5–107°/9 mm	42
70	indefinite	120–120.5°/9 mm	42
1.65	very toxic	136–136.5°/11 mm	42
> 100	non-toxic	145–146°/9 mm	42
7	toxic	*ca.* 135°/0.9 mm	42
18	toxic	143–148°/0.6 mm	42
2.5	very toxic	103–105°/14 mm	8
7	toxic	128–130°/13 mm	8
10	toxic	145–149°/12 mm	8

References p. 108

TABLE XI *(continued)*

Compound	Formula
(d) *Acid derivatives*	
3-Fluoropropionic anhydride	$[F(CH_2)_2CO]_2O$
6-Fluorohexanoamide	$F(CH_2)_5CONH_2$
(e) *Unsaturated acids and esters*	
Ethyl 4-fluorocrotonate	$FCH_2CH{=}CHCOOC_2H_5$
12-Fluorododec-2-enoic acid	$F(CH_2)_9CH{=}CHCOOH$
13-Fluorotridec-2-enoic acid	$F(CH_2)_{10}CH{=}CHCOOH$
(f) *Hydroxyacids and esters*	
Ethyl ω-fluorolactate	$FCH_2CH(OH)COOC_2H_5$
Methyl 4-fluoro-3-hydroxybutyrate	$FCH_2CH(OH)CH_2COOCH_3$
Fluorocitric acid	$HOOCCHFC(OH)(COOH)CH_2COOH$
Dimethyl fluoromalate	$CH_3OOCCHFCH(OH)COOCH_3$
(g) *Ketoacids and esters*	
Fluoropyruvic acid	$FCH_2COCOOH$
Diethyl fluoro-oxalacetate	$C_2H_5OOCCHFCOCOOC_2H_5$
Ethyl ω-fluoroacetoacetate	$FCH_2COCH_2COOC_2H_5$
Ethyl 8-fluoro-3-oxo-octanoate	$F(CH_2)_5COCH_2COOC_2H_5$
Ethyl 9-fluoro-3-oxononanoate	$F(CH_2)_6COCH_2COOC_2H_5$
Ethyl 12-fluoro-3-oxododecanoate	$F(CH_2)_9COCH_2COOC_2H_5$
(h) *Miscellaneous compounds* (included for comparison with the above)	
9(10)-Fluorostearic acid	$CH_3(CH_2)_7CHF(CH_2)_8COOH$
10-Chlorodecanoic acid	$Cl(CH_2)_9COOH$
Monofluorosuccinic acid	$HOOCCHFCH_2COOH$

* See remarks relating to Tables (p. 61).
** See footnote (p. 82).

ω-FLUOROCARBOXYLIC ACIDS AND DERIVATIVES

LD_{50} (mice, I.P.) mg/kg	*Conclusion*	*Boiling point*	*References*
137	non-toxic	123–123.5°/14 mm	38
0.7	very toxic	m.p. 78.5–79°	38
1.25	very toxic	53.5–54°/11 mm	42
1.55	very toxic	m.p. 37–37.5°	42
72	indefinite	m.p. 38–39°	42
> 100	non-toxic	74°/12 mm	44
—	very toxic	98°/19 mm	10, 27, 51
—	indefinite	hygr. solid	17, 54
—	non-toxic	90°/0.5 mm	48, 56
80	non-toxic	98°/5 mm	29
> 400	non-toxic	120–122°/9 mm	6
2.5	very toxic	78–80°/12 mm	13
1.3	very toxic	107°/3 mm	13
67	indefinite	122°/1.7 mm	13
1.9	very toxic	m.p. 47–47.5°	13
> 400	non-toxic	m.p. 68–69°	46
68.8	indefinite	m.p. 39–39.5°	42, 46
—	non-toxic	—	20

References p. 108

REFERENCES

1. AVI-DOR, Y. and MAGER, J. (1956) A spectrophotometric method for determination of cysteine and related compounds. *J. Biol. Chem., 222:* 249.
2. AVI-DOR, Y., and MAGER, J. (1956) The effect of fluoropyruvate on the respiration of animal-tissue preparations. *Biochem. J., 63:* 613.
3. BEINERT, H., BOCK, R. M., GOLDMAN, D. S., GREEN, D. E., MAHLER, H. R., MII, S., STANSLY, P. G., and WAKIL, S. J. (1953) The reconstruction of the fatty acid oxidizing system of animal tissues. *J. Am. Chem. Soc., 75:* 4111.
4. BEINERT, H., GREEN, D. E., HELE, P., HIFT, H., VON KORFF, R. W., and RAMAKRISHNAN, C. V. (1953) The acetate activating enzyme system of heart muscle. *J. Biol. Chem., 203:* 35.
5. BLANK, I., and MAGER, J. (1954) Studies on organic fluorine compounds. II. Esters of oxalofluoroacetic acid. *Experientia, 10:* 77.
6. BLANK, I., MAGER, J., and BERGMANN, E. D. (1955) Studies of organic fluorine compounds. Part IV. Synthesis of esters of fluoro-oxaloacetic and of fluoropyruvic acid. *J. Chem. Soc., 1955:* 2190.
7. BOWMAN, R. E., and FORDHAM, W. D. (1952) Experiments on the synthesis of carbonyl compounds. Part VI. A new general synthesis of ketones and β-keto-esters. *J. Chem. Soc., 1952:* 3945.
8. BUCKLE, F. J., PATTISON, F. L. M., and SAUNDERS, B. C. (1949) Toxic fluorine compounds containing the C-F link. Part VI. ω-Fluorocarboxylic acids and derivatives. *J. Chem. Soc., 1949:* 1471.
9. BUCKLE, F. J., and SAUNDERS, B. C. (1949) Toxic fluorine compounds containing the C-F link. Part VIII. ω-Fluorocarboxylic acids and derivatives containing an oxygen atom as a chain member. *J. Chem. Soc., 1949:* 2774.
10. CHENOWETH, M. B. (1949) Monofluoroacetic acid and related compounds. *J. Pharmacol. Exptl. Therap., II, 97:* 383. *Pharmacol. Revs., 1:* 383.
11. CHENOWETH, M. B. (1958) Personal communication, June 20, 1958.
12. CHENOWETH, M. B., KANDEL, A., JOHNSON, L. B., and BENNETT, D. R. (1951) Factors influencing fluoroacetate poisoning. Practical treatment with glycerol monoacetate. *J. Pharmacol. Exptl. Therap., 102:* 31.

13. FRASER, R. R., MILLINGTON, J. E., and PATTISON, F. L. M. (1957) Toxic fluorine compounds. XV. Some ω-fluoro-β-ketoesters and ω-fluoroketones. *J. Am. Chem. Soc., 79:* 1959.

14. FURCHGOTT, R. F. (1950) The effect of sodium fluoroacetate on the contractility and metabolism of intestinal smooth muscle. *J. Pharmacol. Exptl. Therap., 99:* 1.

15. FURCHGOTT, R. F., and SHORR, E. (1946) Sources of energy for intestinal smooth muscle contraction. *Proc. Soc. Exptl. Biol. Med., 61:* 280.

16. GAL, E. M. (1958) Inhibition of malate and acetoacetate synthesis by their fluoro analogs. *Arch. Biochem. Biophys., 73:* 279.

17. GAL, E. M., PETERS, R. A., and WAKELIN, R. W. (1956) Some effects of synthetic fluoro compounds on the metabolism of acetate and citrate. *Biochem. J., 64:* 161.

18. GITTER, S. (1956) The influence of acetamide on citrate accumulation after fluoroacetate poisoning. *Biochem. J., 63:* 182.

19. GITTER, S., BLANK, I., and BERGMANN, E. D. (1953) Studies on organic fluorine compounds. I. The influence of acetamide on fluoroacetate poisoning. *Koninkl. Ned. Akad. Wetenschap. Proc. Ser. C., 56:* 423.

20. GITTER, S., BLANK, I., and BERGMANN, E. D. (1953) Studies on organic fluorine compounds. II. Toxicology of higher alkyl fluoroacetates. *Koninkl. Ned. Akad. Wetenschap. Proc. Ser. C, 56:* 427.

21. HENDERSHOT, L. C., and CHENOWETH, M. B. (1954) Antagonism of three homologous monofluorinated fatty acids by several fatty acids. *J. Pharmacol. Exptl. Therap., 110:* 344.

22. HENDERSHOT, L. C., and CHENOWETH, M. B. (1955) Fluoroacetate and fluorobutyrate convulsions in the isolated cerebral cortex of the dog. *J. Pharmacol. Exptl. Therap., 113:* 160.

23. HOWELL, W. C., COTT, W. J., and PATTISON, F. L. M. (1957) Organometallic reactions of ω-fluoroalkyl halides. II. Reactions of ω-fluoroalkylmagnesium chlorides. *J. Org. Chem., 22:* 255.

24. KALNITSKY, G., and BARRON, E. S. G. (1948) The inhibition by fluoroacetate and fluorobutyrate of fatty acid and glucose oxidation produced by kidney homogenates. *Arch. Biochem., 19:* 75.

25. KANDEL, A., and CHENOWETH, M. B. (1952) Metabolic disturbances produced by some fluoro-fatty acids: relation to the pharmacologic

activity of these compounds. *J. Pharmacol. Exptl. Therap.*, *104:* 234.

26. KANDEL, A., and CHENOWETH, M. B. (1952) Tolerance to fluoroacetate and fluorobutyrate in rats. *J. Pharmacol. Exptl. Therap.*, *104:* 248.
27. KHARASCH, M. S., *et al.* (1943–1945) Unpublished results.
28. KNOOP, F. (1904) Der Abbau aromatischer Fettsäuren im Tierkörper. *Beitr. chem. Physiol. Pathol.*, *6:* 150.
29. MAGER, J. and BLANK, I. (1954) Synthesis of fluoropyruvic acid and some of its biological properties. *Nature*, *173:* 126.
30. MAGER, J., GOLDBLUM-SINAI, J., and BLANK, I. (1955) Effect of fluoroacetic acid and allied fluoroanalogues on growth of *Escherichia coli.* I. Pattern of inhibition. *J. Bacteriol.*, *70:* 320.
31. MAHLER, H. R., WAKIL, S. J., and BOCK, R. M. (1953) Studies on fatty acid oxidation. I. Enzymic activation of fatty acids. *J. Biol. Chem.*, *204:* 453.
32. MAVITY, J. M. (1957) β-Fluorocarboxylic acids. *U.S. Patent 2,798, 091*, July 2, 1957.
33. NAIR, P. V., and BUSCH, H. (1958) Improved synthesis of monofluoro- and monochloropyruvic acids. *J. Org. Chem.*, *23:* 137.
34. NISCHK, G., and MÜLLER, E. (1952) Bemerkungen zu den Arbeiten von R. Huisgen und J. Reinertshofer: Nitroso-acyl-amine und Diazoester VII und VIII. *Ann.*, *576:* 232.
35. OLIVER, DANIEL (1868) *Flora of tropical Africa*, Vol. I, London, p. 341.
36. PARKER, J. M., and WALKER, I. G. (1957) A toxicological and biochemical study of ω-fluoro compounds. *Can. J. Biochem. Physiol.*, *35:* 407.
37. PATTISON, F. L. M. (1954) Toxic fluorine compounds. II. *Nature*, *174:* 737.
38. PATTISON, F. L. M., FRASER, R. R., O'NEILL, G. J., and WILSHIRE, J. F. K. (1956) Toxic fluorine compounds. X. ω-Fluorocarboxylic acid chlorides, anhydrides, amides and anilides. *J. Org. Chem.*, *21:* 887.
39. PATTISON, F. L. M., and HOWELL, W. C. (1956) Organometallic reactions of ω-fluoroalkyl halides. I. Preparation of ω-fluoroalkylmagnesium halides. *J. Org. Chem.*, *21:* 879.

40. PATTISON, F. L. M., HOWELL, W. C., McNAMARA, A. J., SCHNEIDER, J. C., and WALKER, J. F. (1956) Toxic fluorine compounds. III. ω-Fluoroalcohols. *J. Org. Chem., 21:* 739.

41. PATTISON, F. L. M., HOWELL, W. C., and WOOLFORD, R. G. (1957) Toxic fluorine compounds. XIII. ω-Fluoroalkyl ethers. *Can. J. Chem., 35:* 141.

42. PATTISON, F. L. M., HUNT, S. B. D., and STOTHERS, J. B. (1956) Toxic fluorine compounds. IX. ω-Fluorocarboxylic esters and acids. *J. Org. Chem., 21:* 883.

43. PATTISON, F. L. M., and MILLINGTON, J. E. (1956) The preparation and some cleavage reactions of alkyl and substituted alkyl methanesulphonates. *Can. J. Chem., 34:* 757.

44. PATTISON, F. L. M., and NORMAN, J. J. (1957) Toxic fluorine compounds. XVII. Some l-fluoroalkanes, ω-fluoroalkenes and ω-fluoroalkynes. *J. Am. Chem. Soc., 79:* 2311.

45. PATTISON, F. L. M., and SAUNDERS, B. C. (1949) Toxic fluorine compounds containing the C-F link. Part VII. Evidence for the β-oxidation of ω-fluorocarboxylic acids *in vivo. J. Chem. Soc., 1949:* 2745.

46. PATTISON, F. L. M., STOTHERS, J. B., and WOOLFORD, R. G. (1956) Anodic syntheses involving ω-monohalocarboxylic acids. *J. Am. Chem. Soc., 78:* 2255.

47. PATTISON, F. L. M., and WOOLFORD, R. G. (1957) Toxic fluorine compounds. XVI. Branched ω-fluorocarboxylic acids. *J. Am. Chem. Soc., 79:* 2308.

48. PETERS, R. A. (1957) Mechanism of the toxicity of the active constituent of *Dichapetalum cymosum* and related compounds. *Advances in Enzymology and Related Subjects of Biochemistry*, Vol. XVIII, Interscience Publishers, Inc., New York, p. 113.

49. PETERS, R. A., and HALL, R. J. (1957) Note upon the reaction of fluoropyruvate with some -SH compounds. *Biochim. Biophys. Acta, 26:* 433.

50. POWER, F. B., and TUTIN, F. (1906) Chemical and physiological examination of the fruit of *Chailletia toxicaria. J. Am. Chem. Soc., 28:* 1170.

51. REDEMANN, C. E., CHAIKIN, S. W., FEARING, R. B., ROTARIU, G. J., SAVIT, J., and VAN HOESEN, D. (1948) The vapor pressures of forty-

one fluorine-containing organic compounds. *J. Am. Chem. Soc., 70:* 3604.

52. RENNER, W. (1904) A case of poisoning by the fruit of *Chailletia toxicaria* (ratsbane). *Brit. Med. J.,* 1314.
53. RENNER, W. (1904–1905) Native poison, West Africa. *J. African Soc., 4:* 109.
54. RIVETT, D. E. A. (1953) The synthesis of monofluorocitric acid. *J. Chem. Soc., 1953:* 3710.
55. TAYLOR, N. F., and KENT, P. W. (1954) Synthesis of (±) methyl fluoromalate. *Nature, 174:* 401.
56. TAYLOR, N. F., and KENT, P. W. (1956) The synthesis of 2-deoxy-2-fluorotetritols and 2-deoxy-2-fluoro-(±)-glyceraldehyde. *J. Chem. Soc., 1956:* 2150.
57. WALKER, I. G., and PARKER, J. M. (1958) Further toxicological and biochemical studies on ω-fluoro compounds. *Can. J. Biochem. Physiol., 36:* 339.
58. WARD, P. F. V., GAL, E. M., and PETERS, R. A. (1956) Note on the preparation of the free acid and of the trisodium salt from the triester of fluorocitric acid. *Biochim. Biophys. Acta, 21:* 392.
59. WARD, P. F. V., and PETERS, R. A. (1957) Experiments on the purification of fluorocitric acid. *Biochim. Biophys. Acta, 26:* 449.
60. WARD, P. F. V., and PETERS, R. A. (1958) Purification and transformation of synthetic fluorocitric acid. *Abstracts of communications presented at the IVth International Congress of Biochemistry, Vienna,* p. 3, (Section 1, Communication 19).
61. WILSHIRE, J. F. K., and PATTISON, F. L. M. (1956) Toxic fluorine compounds. XI. ω-Fluoroaldehydes. *J. Am. Chem. Soc., 78:* 4996.

4

Other ω-fluoro compounds

This chapter contains a description of various homologous series which show an alternation of toxic properties but which have not been described earlier in the monograph. Most of the compounds have in common the presence of an ω-fluorine atom attached to an unbranched carbon chain. The two-carbon members of a few of the series have already been discussed; thus, 2-fluoroethanol and fluoroacetaldehyde were considered along with the fluoroacetates in Chapter 2. The arbitrary method of selection is outlined on p. 12.

It should be mentioned at this point that all the toxic members of the series described below gave rise to typical fluoroacetate-like symptoms *in vivo*. Moreover, in the limited number of compounds examined, Parker and Walker observed[45] that the toxic members resulted in a pronounced accumulation of citric acid, whereas the non-toxic members caused none. Both these facts may be taken as indirect evidence that *in a wide variety of compounds of general formula $F(CH_2)_nZ$, the grouping Z is first metabolized by one of the natural detoxication mechanisms of the body, and that the resulting ω-fluorocarboxylate then undergoes β-oxidation* (as outlined earlier, p. 90). Final proof of this interpretation as applied to the toxic members will of course require the isolation and identification of fluoroacetate* or fluorocitrate in the tissues of poisoned animals.

---

* The end product in the β-oxidation process is thought to be fluoroacetyl-coenzyme A, but for convenience the term 'fluoroacetate' is usually employed in the discussions. It is not necessary to postulate the formation of free fluoroacetic acid.

No toxicity results with humans have been recorded for any of the compounds described in this chapter.

As the work which led up to this conclusion was proceeding, it gradually became apparent[47] that the ω-fluorine atom might be of potential value as a 'tag' for elucidating intermediary metabolism. The method, which is discussed with examples in Chapter 5 (p. 182), developed in two stages: early in the work it was found to be possible to explain toxicity patterns of some of the series in terms of known metabolic detoxication processes; this later gave rise to the natural corollary that an *unknown* metabolic process could be elucidated by examining the toxicity pattern of the ω-fluoro series containing the particular grouping under consideration.

The majority of compounds to be discussed are stable, colourless liquids, with odours similar to those of the non-fluorinated materials. The effect of the fluorine atom on melting points is unpredictable, and little can be said about any rules of thumb relating those of ω-fluoro compounds to those of the other ω-halo compounds or non-halogenated analogues. It is possible to be more specific about boiling points: as a rough guide, the boiling point (referred to atmospheric pressure) of most ω-fluoro compounds excepting the very low members is often about 25° to 35° lower than that of the corresponding ω-chloro compound, about 40° to 50° lower than that of the corresponding ω-bromo compound, and about 35° to 45° higher than that of the corresponding non-halogenated analogue.

FLUORINATED HYDROCARBONS

1-Fluoroalkanes, $CH_3(CH_2)_nF$

The 1-fluoroalkanes may be prepared conveniently from alkyl halides or alkyl sulphonates by treatment with potassium fluoride in diethylene glycol[6, 13, 21, 23, 55, 56, 73]. Detailed instructions for the preparation of 1-fluorohexane from 1-hexanol by method (b) are given in Appendix II.

(a) $CH_3(CH_2)_nX + KF \rightarrow CH_3(CH_2)_nF + KX \qquad (X = Cl \text{ or } Br)$

(b) $CH_3(CH_2)_nOSO_2R + KF \rightarrow CH_3(CH_2)_nF + RSO_2OK$

They are colourless, very stable liquids. Toxicity results[56] and physical constants are listed in Table XII. The most striking feature of these figures is the very high toxicity of the 1-fluoroalkanes, and more particularly of the members containing an even number of carbon atoms. They are indeed among the most toxic of the ω-fluoro compounds; this fact, coupled with their relatively high volatility, emphasizes the need for caution in their manipulation. The hazardous nature of the 1-fluoroalkanes is the more alarming and surprising when contrasted with the outstanding stability of the members, and points a warning to industrial workers who may encounter these materials as by-products in commercial processes involving the use of fluorine or its compounds.

TABLE XII

1-FLUOROALKANES

Compound	*Formula*	*LD_{50} (mice, I.P.) mg/kg*	*Conclusion*	*Boiling point*
1-Fluorohexane	$CH_3(CH_2)_5F$	1.7	very toxic	91–92°
1-Fluoroheptane	$CH_3(CH_2)_6F$	35	indefinite	119–121°
1-Fluoro-octane	$CH_3(CH_2)_7F$	2.7	very toxic	144–146°
1-Fluorononane	$CH_3(CH_2)_8F$	21.7	toxic	166–169°
1-Fluorodecane	$CH_3(CH_2)_9F$	1.7	very toxic	186–188°
1-Fluoroundecane	$CH_3(CH_2)_{10}F$	15.5	toxic	70–71.5°/3mm
1-Fluorododecane	$CH_3(CH_2)_{11}F$	2.5	very toxic	93–95°/3mm

The alternation in toxicity of the 1-fluoroalkanes clearly shown in Table XII indicates the existence of a biochemical degradative mechanism akin to that of the ω-fluorocarboxylates (p. 5). A reasonable interpretation of this observation consequently points to initial ω-oxidation of the 1-fluoroalkanes, resulting in the formation of ω-fluorocarboxylates containing the same number of carbon atoms.

$$F(CH_2)_nCH_3 \rightarrow F(CH_2)_nCOOH$$

The fluoroacids, once formed, would then be degraded by β-oxidation (p. 90). In support of this is the marked similarity in toxicity between the 1-fluoroalkanes, particularly those containing an even number of carbon atoms, and the corresponding ω-fluorocarboxylates; for example, the toxicities (mg/kg) to mice of the six- and seven-carbon members of the two series are as follows:

$F(CH_2)_5CH_3$	1.7	$F(CH_2)_5COOH$	1.35
$F(CH_2)_6CH_3$	35	$F(CH_2)_6COOH$	40

Further evidence in favour of the ω-oxidation mechanism is supplied by high citric acid levels found in animals poisoned by the even 1-fluoroalkanes[75], indicating (p. 93) the ultimate formation of fluoroacetate. It should be mentioned however that the odd 1-fluoroalkanes also promoted citric acid accumulation, but more slowly[75].

The relatively high toxicity of the *odd* 1-fluoroalkanes indicates that ω-oxidation is not the sole factor concerned with their metabolism. To explain this, Walker and Parker[75] have suggested that oxidative attack on the 1-fluoroalkanes may occur not only at the end but also at some non-terminal point of the chain, with subsequent chain cleavage. From studies of citric acid accumulation, it was concluded however that ω-oxidation was the predominant effect, with chain cleavage playing a significant but subsidiary rôle.

ω,ω′-Difluoroalkanes, $F(CH_2)_nF$

The ω,ω′-difluoroalkanes may be prepared[22, 23, 51, 57] either by total halogen exchange using the appropriate ω,ω′-dihaloalkanes or by symmetrical anodic coupling of ω-fluorocarboxylic acids.

(a) $X(CH_2)_nX + 2KF \rightarrow F(CH_2)_nF + 2KX$

(b) $2\,F(CH_2)_nCOOH \rightarrow F(CH_2)_{2n}F + 2CO_2 + H_2$

Because of the resistance of the carbon-fluorine bond to hydrolysis, it was expected that these compounds would be excreted unchanged. The members listed in Table XIII were submitted for routine testing and, surprisingly, were found to be very toxic[51]; in contrast, 1,18-dichloro-octadecane, $Cl(CH_2)_{18}Cl$ was non-toxic[51].

Walker and Parker[45, 75] have studied these compounds in regard to their ability to promote citric acid accumulation. It was shown that 1,10-difluorodecane in small dosage resulted in massive accumulation of citric acid, whereas 1,7-difluoroheptane in large dosage resulted in none[45]. In contrast to the conclusions relating to the 1-fluoroalkanes outlined above, it is likely that oxidative chain cleavage is the operative effect:

$$F(CH_2)_nCH_2CH_2(CH_2)_mF \rightarrow F(CH_2)_nCOOH + F(CH_2)_mCOOH$$

ω-Fluoroalkenes, $F(CH_2)_nCH=CH_2$

The ω-fluoroalkenes were prepared from ω-haloalkenes[56] or from ω-methanesulphonoxyalkenes[55] by treatment with potassium fluoride in diethylene glycol.

(a) $X(CH_2)_nCH=CH_2 + KF \rightarrow F(CH_2)_nCH=CH_2 + KX$
(b) $CH_3SO_2O(CH_2)_nCH=CH_2 + KF \rightarrow F(CH_2)_nCH=CH_2 + CH_3SO_2OK$

The four members listed in Table XIV are all toxic[56], irrespective of the length of their carbon chains. Hence no clear-cut conclusions can be drawn from these figures regarding the biochemical fate of the alkene grouping. It should be noted that certain polyfluoroalkenes are also toxic (see p. 125).

ω-Fluoroalkynes, $F(CH_2)_nC\equiv CH$

The ω-fluoroalkynes[56] were prepared either by treatment of the appropriate ω-fluoroalkyl bromides and iodides with sodium acetylide, or by halogen exchange using the corresponding ω-haloalkynes:

(a) $F(CH_2)_nX + NaC\equiv CH \rightarrow F(CH_2)_nC\equiv CH + NaX$
(b) $X(CH_2)_nC\equiv CH + KF \rightarrow F(CH_2)_nC\equiv CH + KX$

The alternation in toxicity of the members as listed in Table XV indicates that the alkyne grouping is attacked *in vivo*. From a comparison of the toxicity pattern with that described below (p. 135) for the ω-fluoroalkyl methyl ketones, it has been suggested[56]

TABLE XIII

Compound	*Formula*
1,4-Difluorobutane	$F(CH_2)_4F$
1,5-Difluoropentane	$F(CH_2)_5F$
1,7-Difluoroheptane	$F(CH_2)_7F$
1,8-Difluoro-octane	$F(CH_2)_8F$
1,10-Difluorodecane	$F(CH_2)_{10}F$
1,12-Difluorododecane	$F(CH_2)_{12}F$
1,14-Difluorotetradecane	$F(CH_2)_{14}F$
1,16-Difluorohexadecane	$F(CH_2)_{16}F$
1,18-Difluoro-octadecane	$F(CH_2)_{18}F$
1,20-Difluoroeicosane	$F(CH_2)_{20}F$
cf. 1,18-Dichloro-octadecane	$Cl(CH_2)_{18}Cl$

TABLE XIV

Compound	*Formula*
5-Fluoro-1-pentene	$F(CH_2)_3CH{=}CH_2$
6-Fluoro-1-hexene	$F(CH_2)_4CH{=}CH_2$
11-Fluoro-1-undecene	$F(CH_2)_9CH{=}CH_2$
1,4-Difluoro-2-butene	$FCH_2CH{=}CHCH_2F$

TABLE XV

Compound	*Formula*
6-Fluoro-1-hexyne	$F(CH_2)_4C{\equiv}CH$
7-Fluoro-1-heptyne	$F(CH_2)_5C{\equiv}CH$
8-Fluoro-1-octyne	$F(CH_2)_6C{\equiv}CH$
9-Fluoro-1-nonyne	$F(CH_2)_7C{\equiv}CH$
cf. 1-Hexyne	$CH_3(CH_2)_3C{\equiv}CH$

ω, ω'-DIFLUOROALKANES

LD_{50} *(mice, I.P.) mg/kg*	*Conclusion*	*Boiling point*
3.4	toxic	77.8°
18	toxic	105.5°
21.3	toxic	48°/10 mm
1.6	very toxic	75–75.5°/13 mm
2.1	very toxic	98°/12 mm
2.5	very toxic	120°/10 mm
2.3	very toxic	148°/11 mm
10.9	toxic	138°/2.7 mm
10	toxic	163–164°/9 mm
10.2	toxic	m.p. 46–46.5°
> 100	non-toxic	m.p. 53.5–54°

ω-FLUOROALKENES

LD_{50} *(mice, I.P.) mg/kg*	*Conclusion*	*Boiling point*
5.4	toxic	61–62°
2.8	very toxic	91.5°
9.3	toxic	84–85°/11 mm
6.1	toxic	73°

ω-FLUOROALKYNES

LD_{50} *(mice, I.P.) mg/kg*	*Conclusion*	*Boiling point*
5.7	toxic	106–106.5°
53	indefinite	131–131.5°
7.5	toxic	77–78°/50 mm
79	non-toxic	66–66.5°/12 mm
> 100	non-toxic	71.5°

that hydration of the alkyne grouping may be the initial step in its degradation:

$$F(CH_2)_nC{\equiv}CH + H_2O \rightarrow F(CH_2)_nCOCH_3$$

The fact that the fluoroalkynes are all rather less toxic than the corresponding fluoroketones may indicate, however, that hydration is not the sole metabolic route. Ultimate formation of fluoroacetate from 6-fluoro-1-hexyne and from 7-fluoro-1-heptyne was shown by citric acid accumulation in the tissues of poisoned animals[44, 45]. That both the even *and* odd members caused citric acid accumulation is a significant point in common between the fluoroalkynes and the fluoroalkyl methyl ketones; moreover, with both classes, the accumulation was greater with the even chain members (on a molar basis). While the toxicity pattern and citric acid levels are thus in substantial agreement with the hydration of the alkyne grouping to the corresponding methyl ketone, independent work is still required to confirm or disprove this postulate.

ω-Fluoroalkyl halides*, $F(CH_2)_nX$

ω-Fluoroalkyl chlorides, bromides and iodides may be obtained by many different methods, most of which have been reviewed in a recent publication[51]. Since some members are extremely toxic, it is perhaps appropriate that these reactions be listed; full details and complete literature references may be obtained from the above-mentioned review. It is the author's opinion that the ω-fluoroalkyl halides are amongst the compounds most likely to be encountered as highly dangerous by-products or contaminants in commercial processes; other previously-mentioned classes which are likely to be particularly hazardous for the same reason include the 1-fluoroalkanes and ω,ω′-difluoroalkanes *(q.v.)*. It is not difficult to envisage how toxic members of these three classes could be formed in appreciable quantities during fluorination and halogenation processes involving hydrocarbons. Most of the known preparations

---

* To avoid ambiguity, fluorine is not generally referred to as halogen in this monograph.

of ω-fluoroalkyl halides may be summarized by the following reactions. Detailed instructions for the preparation of 6-fluorohexyl chloride by method (a) are given in Appendix II.

(a) $X(CH_2)_nX + MF \rightarrow F(CH_2)_nX + MF \quad (M = K, Hg^+, \frac{1}{2} Hg^{++})$
(b) $3\ F(CH_2)_nOH + PX_3 \rightarrow 3\ F(CH_2)_nX + H_3PO_3$
(c) $F(CH_2)_nOH + SOCl_2 \rightarrow F(CH_2)_nCl + SO_2 + HCl$
(d) $F(CH_2)_nOSO_2R + KX \rightarrow F(CH_2)_nX + RSO_2OK$
(e) $X(CH_2)_nOSO_2R + KF \rightarrow X(CH_2)_nF + RSO_2OK$
(f) $X(CH_2)_nOSO_2OK + KF \rightarrow X(CH_2)_nF + K_2SO_4$
(g) $F_2 + X_2 \rightarrow 2\ FX;\ FX + CH_2{=}CH_2 \rightarrow FCH_2CH_2X$
(h) $F(CH_2)_nCOOH + HOOC(CH_2)_mX \rightarrow F(CH_2)_{n+m}X + 2\ CO_2 + H_2$

Chemically, the main value of these derivatives lies in the greater stability of the carbon-fluorine bond relative to that of the other carbon-halogen bonds, making it possible to replace the halogen atom by standard reactions without affecting the fluorine atom[51]. They have thus became valuable intermediates in the synthesis of ω-fluoro compounds. This aspect of the work is outlined briefly in Appendix I.

2-Fluoroethyl chloride and bromide were found to be relatively non-toxic[61]; this fact was correlated with the unreactive nature of the halogen atom in these short-chain members, and hence of their inability to be hydrolyzed *in vivo* to the toxic 2-fluoroethanol[61]. From more recent work, however, it became apparent that the halogen of the higher ω-fluoroalkyl halides is more labile than that of the 2-fluoroethyl halides. This suggested that the higher members might more readily be hydrolyzed *in vivo* to the corresponding ω-fluoroalcohols. Thus it was predicted[51] that the even members might be more toxic than the odd members. This is confirmed[51] by the figures shown in Table XVI. (It can be seen that 2-fluoroethyl iodide is apparently more susceptible to hydrolysis than the other 2-fluoroethyl halides.)

$$F(CH_2)_nX + H_2O \rightarrow F(CH_2)_nOH + HX$$

The chlorides and iodides seem to be rather more toxic than the bromides, but the typical alternation is evident on ascent of all three series. The toxicology of all but the lowest members thus conforms closely to that of the ω-fluoroalcohol series (p. 127) and this in turn confirms the postulated hydrolytic removal of the halogen atom. Supporting this conclusion are some studies of citric acid accumulation[45]: 6-fluorohexyl chloride, bromide and iodide all gave rise to large accumulations of citric acid in mice, proving that fluoroacetate had been formed, presumably by hydrolytic dehalogenation, oxidation and β-oxidation:

$$F(CH_2)_6X \rightarrow F(CH_2)_6OH \rightarrow F(CH_2)_5COOH \rightarrow FCH_2COOH$$

TABLE XVI

Compound	*Formula*
2-Fluoroethyl chloride	$F(CH_2)_2Cl$
2-Fluoroethyl bromide	$F(CH_2)_2Br$
2-Fluoroethyl iodide	$F(CH_2)_2I$
3-Fluoropropyl bromide	$F(CH_2)_3Br$
4-Fluorobutyl chloride	$F(CH_2)_4Cl$
4-Fluorobutyl bromide	$F(CH_2)_4Br$
4-Fluorobutyl iodide	$F(CH_2)_4I$
5-Fluoroamyl chloride	$F(CH_2)_5Cl$
5-Fluoroamyl bromide	$F(CH_2)_5Br$
5-Fluoroamyl iodide	$F(CH_2)_5I$
6-Fluorohexyl chloride	$F(CH_2)_6Cl$
6-Fluorohexyl bromide	$F(CH_2)_6Br$
6-Fluorohexyl iodide	$F(CH_2)_6I$
7-Fluoroheptyl chloride	$F(CH_2)_7Cl$
7-Fluoroheptyl bromide	$F(CH_2)_7Br$
8-Fluoro-octyl chloride	$F(CH_2)_8Cl$
8-Fluoro-octyl bromide	$F(CH_2)_8Br$
9-Fluorononyl chloride	$F(CH_2)_9Cl$
10-Fluorodecyl chloride	$F(CH_2)_{10}Cl$
10-Fluorodecyl bromide	$F(CH_2)_{10}Br$
11-Fluoroundecyl bromide	$F(CH_2)_{11}Br$
12-Fluorododecyl bromide	$F(CH_2)_{12}Br$
13-Fluorotridecyl chloride	$F(CH_2)_{13}Cl$

Walker and Parker[75] have investigated some of the *odd* ω-fluoroalkyl halides, in particular the five-carbon members, both in regard to their unusually high toxicity and to their ability to promote citrate accumulation. The only reasonable interpretation of their results involves chain cleavage, apparently at a rate slower than that of the hydrolytic dehalogenation postulated above. It is therefore not improbable that the ω-fluoroalkyl halides are degraded by both mechanisms, with chain cleavage playing a subsidiary rôle. Such a conclusion may with justification be extended to non-

ω-FLUOROALKYL HALIDES

LD_{50} *(mice, I.P.) mg/kg*	*Conclusion*	*Boiling point*
> 100	non-toxic	53.5–54°
> 100	non-toxic	70–71°
28	indefinite	89–91°
> 100	non-toxic	100–101°
1.2	very toxic	114–114.5°
8.2	toxic	134–135°
5.2	toxic	52.5–53.5°/13 mm
32	indefinite	143–143.5°
10.5	toxic	54.5–55°/13 mm
8.5	toxic	71–72°/11 mm
5.8	toxic	61.5–62°/15 mm
12.8	toxic	67–68°/11 mm
4.5	toxic	89–89.5°/13 mm
> 100	non-toxic	70–71°/10 mm
> 100	non-toxic	85–86°/11 mm
2.3	very toxic	87–87.5°/10 mm
20	toxic	118–120°/22.5 mm
> 100	non-toxic	102–102.5°/11 mm
5.0	toxic	115–115.5°/9 mm
20	toxic	131–132°/11 mm
> 100	non-toxic	95–96°/0.6 mm
16	toxic	85–86°/0.15 mm
40	indefinite	160–161°/14.5 mm

References p. 160

fluorinated alkyl halides, thus affording indirect evidence for the biological dehalogenation of aliphatic halo compounds.

Polyfluorinated hydrocarbons

Apart from the compounds described above, most saturated fluorinated hydrocarbons are relatively non-toxic. Moreover, most are chemically inert and, if the proportion of hydrogen in the molecule is low, non-inflammable. These properties are particularly important in the case of industrial products such as: (a) the chlorofluorocarbons, such as dichlorodifluoromethane, CF_2Cl_2, which are used extensively as refrigerants and aerosol propellants; (b) the liquid fluorocarbons, C_nF_{2n+2}, which are becoming increasingly important as high-temperature lubricants ('fluoro-greases'); (c) bromofluoro compounds, which are effective fire-extinguishers; (d) plastics, resins, elastomers and rubbers derived from tetrafluoroethylene or chlorotrifluoroethylene (*e.g.* Teflon and Kel-F), which are invaluable for withstanding corrosion or moderately high temperatures (industrial applications include their use as bearings, gaskets, piston rings, valve packings, pipe linings and electrical insulation materials); and (e) miscellaneous polyfluoro compounds which find use as wetting agents, pharmaceutical intermediates, films, sealants, dielectrics, and protective coatings. The inhalation anaesthetic Fluothane (which contains the CF_3-grouping) is described in Chapter 5.

From the available information, the following observations may be made regarding the toxicity of polyfluoro aliphatic compounds. Except for a few polyfluoroalkenes, none can be compared with the ω-monofluoro compounds in terms of potency and speed of action; indeed, oral toxicity at doses as high as 2000 mg/kg/day is negligible[17]. Nevertheless, many are hazardous by inhalation, and can cause extensive damage to the pulmonary system[17, 28, 31]; others can result in narcosis and anaesthesia[17, 31].

(a) Polyfluoroalkanes. Whereas perfluoroalkanes, C_nF_{2n+2} are generally non-toxic[28], chlorofluoroalkanes show variable toxicity[17, 31]. It may be stated that, in the methane and ethane series of completely halogenated compounds, the toxicity decreases with

increasing fluorine content and increases with increasing chlorine content[28]. The members of the ethane series appear to be rather more irritating than those of the methane series[28].

(b) Polyfluoroalkenes. Some of the polyfluoroalkenes are very much more toxic than the polyfluoroalkanes, but few if any reliable generalizations can be drawn from the reported results; thus, it is difficult to reconcile the fairly high toxicity of tetrafluoroethylene, $CF_2{=}CF_2$ with the negligible toxicity of $CF_2{=}CH_2$, 1,1-difluoroethylene [28, 31]; this compound (80%) in oxygen (20%) produced no toxic effects on animals exposed for 19 hours. A serious hazard is presented by the perfluoroalkenes, for example perfluoroisobutylene, $(CF_3)_2C{=}CF_2$ (to which reference is made in the following paragraph) and perfluoropropylene, $CF_3CF{=}CF_2$. The chloropolyfluoroalkenes may also show considerable toxicity; for example, chlorotrifluoroethylene, $CF_2{=}CFCl$ (4 hour exposure at 0.5% by volume) produced no immediate signs of intoxication, but some 18 hours later all the animals (dogs, rabbits and rats) were found dead, pathological changes being confined to pulmonary hypostatic congestion and oedema[28]. On the other hand, the closely related 1,1-dichloro-2,2-difluoroethylene, $CF_2{=}CCl_2$ is less toxic[28]. Until more results are available, it would be wise to exercise extreme caution in working with volatile compounds containing the $CF_2{=}$ grouping.

It is perhaps desirable to add a word of warning about the handling of polymers such as Teflon (polytetrafluoroethylene) and Kel-F (polychlorotrifluoroethylene). Although physiologically inert and non-toxic when ingested or inhaled[1, 78], when heated they become very dangerous[19, 69, 70–72], due to liberation of noxious fumes containing injurious dusts, hydrogen fluoride, and a variety of fluorocarbons including the toxic perfluoroisobutylene, $(CF_3)_2C{=}CF_2$[8, 78]; this last compound is some ten times more toxic than phosgene[78]. On the basis of experiments with animals, Treon and colleagues[69, 71] have shown that fumes from Teflon heated to 500–800° are much more toxic than those formed at 300–500°; indeed, no deaths occurred among animals exposed for five hours to fumes from Teflon heated at 300°. These authors

References p. 160

observed further that the decomposition products formed in an atmosphere of nitrogen are at least as toxic as those formed in air[69, 71], and that the *rapid* destruction of Teflon in either atmosphere yielded products more toxic than those formed more gradually[72]. The following account [71] gives some idea of the effects on animals of the fumes from Teflon heated at 500–800°: 'The fumes [a dense mist, containing finely divided particulate matter] were irritating to the mucous membranes of the animals, there being a prompt outpouring of moisture from eyes, nostrils and mouth. Under the more severe conditions, these discharges became sanguineous. Respiratory distress, dyspnea, gasping and prostration were evident effects, and convulsions also occurred. Even when only 0.92 g of Teflon was decomposed [in a 223-litre chamber], signs of discomfort and dyspnea were observed within 9 to 14 minutes after the exposure began. The corneas of the eyes of many of the animals were etched'. Postmortem examination showed acute pulmonary irritation, characterized by haemorrhage and oedema, and diffuse degeneration of the brain, liver and kidneys[20, 69].

In man, the first symptom, occurring after a latent period of a few hours, is usually a sense of discomfort in the chest. The patient may or may not develop a dry irritating cough. Sometimes these preliminary symptoms are absent, in which case the initial indications of illness are a gradual increase in temperature, pulse-rate, and possibly respiration-rate, followed by shivering and sweating ('polymer-fume fever'; 'the shakes'). The patient usually recovers fairly quickly, provided he is removed from contact with the fumes and kept at rest[20, 66]. It has been reported more recently[11] that, during sintering of Teflon at 350–380°, seven cases of fever developed among workers exposed to the fumes; in one instance, the fluorine content of the urine was found to be as high as 5 mg/l[11]. Skin irritations have been observed during spraying operations[11]. In short, it is essential that proper precautions be taken to safeguard personnel handling these materials; for example, in addition to efficient ventilation, it is advisable to prohibit smoking when there is danger of polymer particles contaminating burning cigarettes. Moreover, caution should be exercised in the storage in vulnerable areas of large quantities of Teflon or Kel-F, since in the event of fire copious amounts of toxic fumes could be liberated.

OXYGEN-CONTAINING ω-FLUORO COMPOUNDS

ω-Fluoroalcohols, $F(CH_2)_nOH$ and derivatives

The ω-fluoroalcohols[52] were prepared by four methods: (a) halogen exchange using the corresponding ω-haloalcohols[52]; (b) halogen exchange using the corresponding ω-haloalkyl acetates followed by hydrolysis[52]; (c) reduction by lithium aluminium hydride of ω-fluorocarboxylic acids or esters[52]; and (d) organometallic reactions[24, 49]. Detailed instructions for the preparation of 6-fluorohexanol by method (a) are given in Appendix II.

(a) $Cl(CH_2)_nOH + KF \rightarrow F(CH_2)_nOH + KCl$

(b) $Cl(CH_2)_nOAc \xrightarrow{KF} F(CH_2)_nOAc \xrightarrow{hydr.} F(CH_2)_nOH$

(c) $F(CH_2)_{n-1}COOR \xrightarrow{LiAlH_4} F(CH_2)_nOH$

(d) $F(CH_2)_{n-2}MgX \xrightarrow{(CH_2)_2O} F(CH_2)_nOH$

The toxicity figures of the ω-fluoroalcohols are listed in Table III and represented by bar-graphs in Fig. 2 (p. 8). For convenience, they are shown in Table XVII with the figures for the corresponding ω-fluorocarboxylic acids (p. 82), and in Table XVIII with physical constants. Death was accompanied by typical fluoroacetate-like symptoms. A pronounced alternation in toxicity is apparent with ascent of the series, the compounds containing an even number of carbon atoms being toxic and those with an odd number non-toxic. With few exceptions, the toxicity data show a striking parallelism between the corresponding members of the two series, thus providing confirmation of the well-established biochemical oxidation of alcohols to acids. In an earlier study with the lowest toxic member (2-fluoroethanol), Bartlett[3] showed that it had no direct action on the enzyme systems of various rat tissues *in vitro*, and hence ascribed its toxic effects in animals to the formation of fluoroacetic acid by tissue alcohol dehydrogenase.

TABLE XVII

COMPARISON OF TOXICITIES OF ω-FLUOROALCOHOLS AND ACIDS

Formula of ω-fluoroalcohol	*LD_{50} (mice, I.P.) mg/kg*	*Formula of corresponding acid*	*LD_{50} (mice, I.P.) mg/kg*
FCH_2CH_2OH	10	FCH_2COOH	6.6
$F(CH_2)_3OH$	46.5	$F(CH_2)_2COOH$	60
$F(CH_2)_4OH$	0.9	$F(CH_2)_3COONa$	0.65
$F(CH_2)_5OH$	> 100	$F(CH_2)_4COOH$	> 100
$F(CH_2)_6OH$	1.2	$F(CH_2)_5COOH$	1.35
$F(CH_2)_7OH$	80	$F(CH_2)_6COOH$	40
$F(CH_2)_8OH$	0.6	$F(CH_2)_7COOH$	0.64
$F(CH_2)_9OH$	32	$F(CH_2)_8COOH$	> 100
$F(CH_2)_{10}OH$	1.0	$F(CH_2)_9COOH$	1.5
$F(CH_2)_{11}OH$	> 100	$F(CH_2)_{10}COOH$	57.5
$F(CH_2)_{12}OH$	1.5	$F(CH_2)_{11}COOH$	1.25
$F(CH_2)_{18}OH$	4.0	$F(CH_2)_{17}COOH$	5.7

TABLE XVIII

Compound	*Formula*
2-Fluoroethanol	FCH_2CH_2OH
3-Fluoropropanol	$F(CH_2)_3OH$
4-Fluorobutanol	$F(CH_2)_4OH$
5-Fluoropentanol	$F(CH_2)_5OH$
6-Fluorohexanol	$F(CH_2)_6OH$
7-Fluoroheptanol	$F(CH_2)_7OH$
8-Fluoro-octanol	$F(CH_2)_8OH$
9-Fluorononanol	$F(CH_2)_9OH$
10-Fluorodecanol	$F(CH_2)_{10}OH$
11-Fluoroundecanol	$F(CH_2)_{11}OH$
12-Fluorododecanol	$F(CH_2)_{12}OH$
18-Fluoro-octadecanol	$F(CH_2)_{18}OH$
cf. 4-Chlorobutanol	$Cl(CH_2)_4OH$
cf. 6-Chlorohexanol	$Cl(CH_2)_6OH$

Reported derivatives of the ω-fluoroalcohols include acetates[34, 52], benzoates[34], sulphonates[34], nitrates[48], and urethanes[40–42, 52, 62, 64]. It is obvious that such toxicity results as are available (Table XIX) conform closely to the pattern described above for the parent ω-fluoroalcohols, thus indicating that most of these derivatives are hydrolyzed readily in the animal body. This is confirmed by the large accumulation of citric acid caused by the administration of 4-fluorobutyl methanesulphonate or *p*-toluenesulphonate[45]; this accumulation presumably results from hydrolysis followed by oxidation of the fluoroalcohol and then β-oxidation to fluoroacetate. It was noted in Chapter 2 (p. 29) that 2-fluoroethyl methanesulphonate and *p*-toluenesulphonate were non-toxic; since 2-fluoroethanol is toxic, the conclusion may be drawn that the two-carbon sulphonate esters are peculiarly and unexpectedly resistant to hydrolysis *in vivo*.

In a recent study, O'Brien and Peters[36] have shown that 2-deoxy-2-fluoro-DL-glyceraldehyde (2-fluoro-3-hydroxypropanal),

ω-FLUOROALCOHOLS

LD_{50} (mice, I.P.) mg/kg	*Conclusion*	*Boiling point*
10	toxic	100–101°
46.5	indefinite	126–127°
0.9	very toxic	57.5–58°/15 mm
$>$ 100	non-toxic	70–71°/11 mm
1.2	very toxic	85–86°/14 mm
80	non-toxic	98–99°/12 mm
0.6	very toxic	106–107°/10 mm
32	indefinite	125–126°/15 mm
1.0	very toxic	136–137°/15 mm
$>$ 100	non-toxic	118–120°/3 mm
1.5	very toxic	88–92°/0.15 mm
4.0	toxic	140–141°/0.01 mm
$>$ 100	non-toxic	81–82°/14 mm
$>$ 100	non-toxic	107–108°/12 mm

$HOCH_2CHFCHO$ and 2-deoxy-2-fluoroglycerol (2-fluoro-1,3-propanediol), $HOCH_2CHFCH_2OH$[67] are toxic to the mouse and the rat, producing convulsions and accumulation of citrate, particularly in the heart and the kidney. A possible explanation of these observations might lie in some oxidative mechanism whereby fluoromalonic acid, once formed, is transformed to fluoroacetic acid by decarboxylation:

$$HOCH_2CHFCHO \rightarrow HOOCCHFCOOH \rightarrow FCH_2COOH + CO_2$$

Results using DL-1-deoxy-1-fluoroglycerol (3-fluoro-1,2-propanediol), $FCH_2CH(OH)CH_2OH$ are mentioned on p.133.

TABLE XIX

Compound	*Formula*
2-Fluoroethyl acetate	$F(CH_2)_2OCOCH_3$
2-Fluoroethyl benzoate	$F(CH_2)_2OCOC_6H_5$
2-Fluoroethyl methanesulphonate	$F(CH_2)_2OSO_2CH_3$
2-Fluoroethyl *p*-toluenesulphonate	$F(CH_2)_2OSO_2C_6H_4CH_3$
3-Fluoropropyl acetate	$F(CH_2)_3OCOCH_3$
3-Fluoropropyl benzoate	$F(CH_2)_3OCOC_6H_5$
3-Fluoropropyl methanesulphonate	$F(CH_2)_3OSO_2CH_3$
3-Fluoropropyl *p*-toluenesulphonate	$F(CH_2)_3OSO_2C_6H_4CH_3$
4-Fluorobutyl acetate	$F(CH_2)_4OCOCH_3$
4-Fluorobutyl benzoate	$F(CH_2)_4OCOC_6H_5$
4-Fluorobutyl methanesulphonate	$F(CH_2)_4OSO_2CH_3$
4-Fluorobutyl *p*-toluenesulphonate	$F(CH_2)_4OSO_2C_6H_4CH_3$
4-Fluorobutyl nitrate	$F(CH_2)_4ONO_2$
5-Fluoroamyl acetate	$F(CH_2)_5OCOCH_3$
5-Fluoroamyl benzoate	$F(CH_2)_5OCOC_6H_5$
5-Fluoroamyl methanesulphonate	$F(CH_2)_5OSO_2CH_3$
5-Fluoroamyl *p*-toluenesulphonate	$F(CH_2)_5OSO_2C_6H_4CH_3$
6-Fluorohexyl methanesulphonate	$F(CH_2)_6OSO_2CH_3$
8-Fluoro-octyl acetate	$F(CH_2)_8OCOCH_3$

ω-Fluoroaldehydes, $F(CH_2)_nCHO$

The members examined[77] were obtained by several different methods.

(a) The Nef reaction

$$F(CH_2)_nCH_2NO_2 \rightarrow F(CH_2)_nCH{=}N(\rightarrow O)ONa \rightarrow F(CH_2)_nCHO$$

(b) The Rosenmund reduction

$$F(CH_2)_nCOCl + H_2 \xrightarrow{Pd} F(CH_2)_nCHO + HCl$$

ESTERS OF ω-FLUOROALCOHOLS

LD_{50} (mice, I.P.) mg/kg	*Conclusion*	*Boiling point*
18.6	toxic	116–117.5°
38	indefinite	112–114°/10 mm
100	non-toxic	118–119°/12 mm
> 100	non-toxic	138.5–140°/1 mm
77.6	non-toxic	41–44°/11 mm
> 100	non-toxic	126–127°/10 mm
> 100	non-toxic	84°/2 mm
> 100	non-toxic	160°/2 mm
0.9	very toxic	56–57°/12 mm
1.7	very toxic	142–143°/10 mm
1.0	very toxic	89–90°/0.5 mm
30	indefinite	151°/1 mm
0.95	very toxic	65°/12 mm
> 100	non-toxic	71–72°/12 mm
> 100	non-toxic	156–157°/11 mm
> 100	non-toxic	112°/2 mm
> 100	non-toxic	163°/0.6 mm
35	indefinite	130–131°/2.5 mm
1.3	very toxic	96–97°/6 mm

(c) The Grignard reaction

$$F(CH_2)_nMgCl \xrightarrow{CH(OEt)_3} F(CH_2)_nCH(OEt)_2 \xrightarrow{hydr.} F(CH_2)_nCHO$$

(d) Periodate oxidation of an ω-fluoro-1,2-alkanediol

$$F(CH_2)_nCH{=}CH_2 \rightarrow F(CH_2)_nCH(OH)CH_2OH \rightarrow F(CH_2)_nCHO$$

The toxicity results presented in Table XX follow the usual dichotomous pattern described for other series of ω-fluoro compounds and in particular for the ω-fluoroalcohols (p. 127) and ω-fluorocarboxylic acids (p. 82). This is understandable in view of the ready oxidation *in vivo* of alcohols and aldehydes to acids. Thus new evidence for the intimate biological relationship of alcohols, aldehydes and acids is furnished, together with yet another example of the ω-fluorine atom acting as a biological tracer. That the even members are ultimately metabolized to fluoroacetate and the odd members are not, is clearly shown[45] by the large accumulation of citric acid following administration of 4-fluorobutanal and the complete absence of accumulation with 9-fluorononanal. The ω-fluoroaldehydes are unstable, undergoing ready polymerization[77].

TABLE XX

Compound	*Formula*
Fluoroacetaldehyde	FCH_2CHO
4-Fluorobutanal	$F(CH_2)_3CHO$
5-Fluoropentanal	$F(CH_2)_4CHO$
6-Fluorohexanal	$F(CH_2)_5CHO$
7-Fluoroheptanal	$F(CH_2)_6CHO$
8-Fluoro-octanal	$F(CH_2)_7CHO$
9-Fluorononanal	$F(CH_2)_8CHO$
10-Fluorodecanal	$F(CH_2)_9CHO$
11-Fluoroundecanal	$F(CH_2)_{10}CHO$

It is of interest that the diols, formed as intermediates in preparative method (d) above, are apparently metabolized to aldehydes or acids[56]. This is supported by the toxicity of 3-fluoro-1,2-propanediol, $FCH_2CH(OH)CH_2OH$ and of 11-fluoro-1,2-undecanediol, $F(CH_2)_9CH(OH)CH_2OH$; for example, the latter has an LD_{50} for mice by intraperitoneal injection of 1.75 mg/kg, which is remarkably similar to that of 10-fluorodecanal (1.95 mg/kg). It seems probable therefore that the α-glycol grouping can be split *in vivo* with loss of one carbon atom to yield first an aldehyde and then an acid. This reasoning, although based on such scanty data as crude toxicity figures, affords the first recorded evidence for the metabolic scission of α-glycols[46]. In a much more thorough independent investigation, O'Brien and Peters[37] have demonstrated recently that 3-fluoro-1,2-propanediol (DL-1-deoxy-1-fluoroglycerol), $FCH_2CH(OH)CH_2OH$ is toxic to the rat, mouse and guinea pig, and that death is associated with convulsions, bradycardia, lowering of body temperature and accumulation of citrate, particularly in the heart and the kidney; in short, an oxidative mechanism involving ultimate scission of the α-glycol linkage is again indicated, resulting in the formation of fluoroacetic acid:

$$FCH_2CH(OH)CH_2OH \rightarrow [\text{intermediate?}] \rightarrow FCH_2COOH$$

ω-FLUOROALDEHYDES

LD_{50} (mice, I.P.) mg/kg	*Conclusion*	*Boiling point*
6.0	toxic	64–65°
2.0	very toxic	49°/50 mm
81	non-toxic	55–56°/20 mm
0.58	very toxic	65–68°/12 mm
> 100	non-toxic	74–75°/10 mm
2.0	very toxic	85–86°/9 mm
53	indefinite	99–100°/9 mm
1.9	very toxic	119–120°/15 mm
> 40	indefinite	130–131°/11 mm

A comparable study of the related 2-fluoro-1,3-propanediol (2-deoxy-2-fluoroglycerol), $HOCH_2CHFCH_2OH$ is described on p. 130.

ω-Fluoroketones, $F(CH_2)_nCOR$

(a) Fluoromethyl ketones, FCH_2COR[15] were prepared from acyl chlorides by conversion to the diazomethyl ketones, followed by treatment with anhydrous hydrogen fluoride[5, 15, 29, 38]; detailed instructions are provided in Appendix II.

$$RCOCl + CH_2N_2 \rightarrow RCOCHN_2 \xrightarrow{HF} RCOCH_2F + N_2$$

This method affords a convenient route to compounds containing the FCH_2CO- grouping. Mono- and difluoroacetone have been obtained by oxidation of the appropriate secondary alcohols[4, 27], and monofluoroacetone by sulphonate cleavage[55] and halogen replacement[27].

The toxicities of the four fluoromethyl alkyl ketones[15, 16] (*i.e.* when R is an unsubstituted alkyl radical) listed in Table XXI indicate that ω-oxidation may play a significant rôle in their metabolism. For example, the non-toxic 1-fluoro-2-heptanone would give rise to a fluoroketoacid which in turn would form the relatively non-toxic fluoropyruvic acid by β-oxidation:

$$FCH_2CO(CH_2)_4CH_3 \rightarrow FCH_2CO(CH_2)_4COOH \rightarrow FCH_2COCOOH$$

On the other hand, the two toxic members by the same mechanism would form the toxic ω-fluoroacetoacetic acid (p. 102):

$$FCH_2CO(CH_2)_7CH_3 \rightarrow FCH_2CO(CH_2)_7COOH \rightarrow FCH_2COCH_2COOH$$

It is possible that direct oxidative cleavage may also occur to a small extent, forming fluoroacetic acid in each instance:

$$FCH_2CO(CH_2)_4CH_3 \rightarrow FCH_2COOH + CH_3(CH_2)_3COOH$$

This would provide a partial explanation for the unexpectedly high toxicity of 1,7-difluoro-2-heptanone, which presumably could also give rise to some of the very toxic 6-fluorohexanoic acid (p. 104) by cleavage between the fluoromethyl and carbonyl groupings. Further evidence for ω-oxidation has been supplied by the results obtained using 1-fluoroalkanes (p. 114), and of oxidative cleavage by the arguments presented below for the ω-fluoroalkyl ketones. As might be expected, tetrafluoroacetone, CHF_2COCHF_2 is relatively non-toxic (LD_{50} *ca.* 250 mg/kg to rats by intravenous injection)[28].

(b) ω-Fluoroalkyl ketones, $F(CH_2)_nCOR$[15]. The mono-and difluorinated aliphatic ketones listed in Table XXI were prepared by Grignard reactions of ω-fluoroalkyl chlorides with the appropriate acid chlorides or anhydrides[24]. The ω-fluoroalkyl phenyl ketones were obtained from benzonitrile by reaction either with ω-fluoroalkyllithium compounds[49] or with ω-fluoroalkylmagnesium halides[24].

$$\text{(a)}\quad F(CH_2)_nMgCl + ClCOR \rightarrow F(CH_2)_nCOR + MgCl_2$$

$$\text{(b)}\quad F(CH_2)_nLi \xrightarrow{C_6H_5CN} F(CH_2)_nCOC_6H_5$$

When considering the compounds of general formula $F(CH_2)_n$-$COCH_3$, it is evident[15] that the members for which n is even are relatively more toxic than those for which n is odd. However, those compounds in which n is odd, although less active, still exhibit toxicities comparable to that of fluoroacetic acid. These facts may be explained by considering two competing modes of breakdown:

$$F(CH_2)_nCOCH_3 \rightarrow F(CH_2)_nCOOH + HCOOH$$
$$F(CH_2)_nCOCH_3 \rightarrow F(CH_2)_{n-1}COOH + CH_3COOH$$

From the toxicity data presented in Table XXI, the first of these is subordinate to the second, although both apparently occur to an appreciable extent. This general trend therefore conforms to the Popoff rule regarding the chemical oxidation of ketones, which states that the carbonyl group normally goes with the smaller radical to a greater extent than with the larger radical. This conclusion may reasonably be extended to include results (p. 88 of ref. 76) obtained using mixed aromatic ketones in which the carbonyl group is not directly attached to the ring. For example:

TABLE XXI

Compound	*Formula*
(a) *Fluoromethyl ketones*	
Fluoroacetone	FCH_2COCH_3
1-Fluoro-2-heptanone	$FCH_2CO(CH_2)_4CH_3$
1-Fluoro-2-octanone	$FCH_2CO(CH_2)_5CH_3$
1-Fluoro-2-decanone	$FCH_2CO(CH_2)_7CH_3$
1,7-Difluoro-2-heptanone	$FCH_2CO(CH_2)_5F$
(b) *ω-Fluoroalkyl ketones*	
8-Fluoro-2-octanone	$F(CH_2)_6COCH_3$
9-Fluoro-2-nonanone	$F(CH_2)_7COCH_3$
10-Fluoro-2-decanone	$F(CH_2)_8COCH_3$
11-Fluoro-2-undecanone	$F(CH_2)_9COCH_3$
12-Fluoro-2-dodecanone	$F(CH_2)_{10}COCH_3$
12-Fluoro-6-dodecanone	$F(CH_2)_6CO(CH_2)_4CH_3$
1,12-Difluoro-6-dodecanone	$F(CH_2)_6CO(CH_2)_5F$
1,13-Difluoro-7-tridecanone	$F(CH_2)_6CO(CH_2)_6F$
1,19-Difluoro-10-nonadecanone	$F(CH_2)_9CO(CH_2)_9F$
ω-Fluoroacetophenone	$FCH_2COC_6H_5$
6-Fluorohexyl phenyl ketone	$F(CH_2)_6COC_6H_5$
8-Fluoro-octyl phenyl ketone	$F(CH_2)_8COC_6H_5$
9-Fluorononyl phenyl ketone	$F(CH_2)_9COC_6H_5$

Ketone	*Predominant end product*
$C_6H_5CH_2COCH_3$	C_6H_5COOH
$C_6H_5CH_2COCH_2CH_2CH_3$	C_6H_5COOH
$C_6H_5CH_2COCH(CH_3)_2$	C_6H_5COOH
$C_6H_5CH_2CH_2COCH_3$	$C_6H_5CH_2COOH$
$C_6H_5CH_2CH_2CH_2CH_2COCH_3$	$C_6H_5CH_2COOH$

Both the odd and even ω-fluoroalkyl methyl ketones gave rise to citric acid accumulation[44, 45], but the effect on a molar basis was greater with the even members, as was expected from the above discussion.

LD_{50} (mice, I.P.) mg/kg	*Conclusion*	*Boiling point*
—	non-toxic	78–79°
60	indefinite	54°/13 mm
8	toxic	70°/11 mm
7.5	toxic	99°/11.5 mm
0.7	very toxic	102–104°/42 mm
3	very toxic	87–88°/11 mm
16	toxic	98–98.5°/9 mm
1.2	very toxic	113–113.5°/9 mm
11.8	toxic	126–126.5°/9.5 mm
1.5	very toxic	138–138.5°/9.5 mm
4.5	toxic	138.5–139°/13 mm
2.2	very toxic	136–136.5°/4 mm
9.6	toxic	99–99.5°/0.18 mm
40	indefinite	m.p. 59°
$>$ 225	non-toxic	90–91°/12 mm
$>$ 100	non-toxic	171–172°/13 mm
100	non-toxic	134–137°/0.7 mm
90	non-toxic	m.p. 36–36.5°

Within the limits of biological variation, the toxicities of 12-fluoro-6-dodecanone, 1,12-difluoro-6-dodecanone and 1,13-difluoro-7-tridecanone appear to confirm the above implication that the predominant metabolic fragments may be predicted correctly by the Popoff rule. Further evidence is supplied by the marked similarity in toxicity of 12-fluoro-6-dodecanone and of 8-fluoro-2-octanone on a molar basis.

The ω-fluoroalkyl phenyl ketones are all relatively non-toxic, possibly due to suppression of the oxidative mechanisms by the phenyl group; these results are in agreement with the low toxicity of the lowest member of the series, ω-fluoroacetophenone, $FCH_2COC_6H_5$, the minimum lethal dose of which has been reported[7] to be 225 mg/kg for rats and mice.

The above arguments, based on simple toxicity determinations and augmented by citric acid studies, may be extended with reservations to unfluorinated ketones. Thus, although the conclusions are tentative and subject to confirmation by more specific procedures, new light has been thrown on the metabolism of ketones in the animal body.

ω-Fluoroalkyl ethers, $F(CH_2)_nOR$

Simple and substituted ω-fluoroethers have been obtained by various means:

(a) By modifications of the Williamson synthesis[54, 61]:

$$F(CH_2)_nOH \rightarrow F(CH_2)_nONa + ICH_3 \rightarrow F(CH_2)_nOCH_3$$
$$F(CH_2)_2Cl + NaOC_6H_5 \rightarrow F(CH_2)_2OC_6H_5 + NaCl$$

(b) By partial or total halogen exchange[54]:

$$Cl(CH_2)_4O(CH_2)_4Cl + KF \rightarrow F(CH_2)_4O(CH_2)_4Cl + KCl$$
$$Cl(CH_2)_4O(CH_2)_4Cl + 2KF \rightarrow F(CH_2)_4O(CH_2)_4F + 2KCl$$

(c) By reaction of sulphonates with potassium fluoride[55]:

$$CH_3SO_2O(CH_2)_2OR + KF \rightarrow F(CH_2)_2OR + CH_3SO_2OK$$

(d) By cyanoethylation of an ω-fluoroalcohol[10]:

$$F(CH_2)_nOH + CH_2{=}CHCN \rightarrow F(CH_2)_nOCH_2CH_2CN$$

(e) By reaction of a fluoroalcohol with ethyl diazoacetate[10]:

$$F(CH_2)_2OH + N_2CHCOOEt \rightarrow F(CH_2)_2OCH_2COOEt + N_2$$

(f) By reaction of a fluoroalcohol with ethylene oxide[10, 63]:

$$F(CH_2)_2OH + (CH_2)_2O \rightarrow F(CH_2)_2O(CH_2)_2OH$$

(g) By reaction of a potassium sulphate ester with potassium fluoride[64]:

$$KOSO_2OCH_2CH_2OCH_2CH_2OH + KF \rightarrow FCH_2CH_2OCH_2CH_2OH + K_2SO_4$$

(h) By special methods, for example[54]:

$$F(CH_2)_nOH + OHCCCl_3 \rightarrow F(CH_2)_nOCH(OH)CCl_3 \xrightarrow{PCl_5} F(CH_2)_nOCHClCCl_3$$

The majority of the toxicity results presented in Table XXII can be explained by simple rupture of the ether link[54]:

$$F(CH_2)_nOR \rightarrow F(CH_2)_nOH$$

Thus, the toxicities of 5-fluoroamyl methyl ether and 6-fluorohexyl methyl ether approximate those of the parent alcohols, 5-fluoropentanol and 6-fluorohexanol; for the same reason, the 3-fluoropropyl ethers, $F(CH_2)_3OR$, which by this degradative process would give rise to 3-fluoropropanol, are non-toxic. Most of the 2-fluoroethyl ethers, FCH_2CH_2OR have molar toxicities in line with this mechanism, within the limits of biological variation. Indeed, Schrader [63, 64] had anticipated this conclusion by observing that 2-fluoroethanol, FCH_2CH_2OH and 2-fluoro-2′-hydroxydiethyl ether, $FCH_2CH_2OCH_2CH_2OH$ were equally toxic. Surprisingly, two of the twelve 2-fluoroethyl ethers (ethyl 2-fluoroethoxyacetate (I) and ethyl 4-(2′-fluoroethoxy)butyrate (II)) have been reported [10] to be non-toxic.

This biochemical cleavage is strikingly confirmed by the results obtained with the four butyl ethers[54]. The dichloro ether (VI), containing no fluorine, is non-toxic. The toxicities of the fluorochloro and fluorocyano ethers (III and IV) are almost identical, and both are very similar to that of 4-fluorobutanol on a molar basis. The difluoro ether (V) can theoretically give rise to twice the quantity of 4-fluorobutanol on hydrolytic fission, and, according

TABLE XXII

Compound	*Formula*
2-Fluoroethyl methyl ether	$FCH_2CH_2OCH_3$
2-Fluoroethyl phenyl ether	$FCH_2CH_2OC_6H_5$
2-Fluoroethyl β-naphthyl ether	$FCH_2CH_2OC_{10}H_7$
2-Fluoro-1′, 2′, 2′, 2′-tetra-chlorodiethyl ether	$FCH_2CH_2OCHClCCl_3$
2-Fluoro-2′-*n*-butoxydiethyl ether	$FCH_2CH_2OCH_2CH_2OBu$
Ethyl 2-fluoroethoxyacetate (I)	$FCH_2CH_2OCH_2COOEt$
2-Fluoro-2′-hydroxydiethyl ether	$FCH_2CH_2OCH_2CH_2OH$
Bis-2-(2′-fluoroethoxy)ethyl methylal	$(FCH_2CH_2OCH_2CH_2O)_2CH_2$
2-Fluoro-2′-(2″-hydroxyethoxy)-diethyl ether	$FCH_2CH_2OCH_2CH_2OCH_2CH_2OH$
2-Fluoro-2′-cyanodiethyl ether	$FCH_2CH_2OCH_2CH_2CN$
3-(2′-Fluoroethoxy)propionic acid	$FCH_2CH_2OCH_2CH_2COOH$
Ethyl 4-(2′-fluoroethoxy)butyrate (II)	$FCH_2CH_2O(CH_2)_3COOEt$
3-(2′-Fluoroethoxy)propylamine	$FCH_2CH_2O(CH_2)_3NH_2$
3-(3′-Fluoropropoxy)propionitrile	$F(CH_2)_3O(CH_2)_2CN$
3-(3′-Fluoropropoxy)propionic acid	$F(CH_2)_3O(CH_2)_2COOH$
Ethyl 4-(3′-fluoropropoxy)butyrate	$F(CH_2)_3O(CH_2)_3COOEt$
3-Fluoropropyl 1′, 2′, 2′, 2′-tetra-chloroethyl ether	$F(CH_2)_3OCHClCCl_3$
4-Fluoro-4′-chlorodibutyl ether (III)	$F(CH_2)_4O(CH_2)_4Cl$
4-Fluoro-4′-cyanodibutyl ether (IV)	$F(CH_2)_4O(CH_2)_4CN$
4,4′-Difluorodibutyl ether (V)	$F(CH_2)_4O(CH_2)_4F$
4-Fluorobutyl 1′, 2′, 2′, 2′-tetra-chloroethyl ether	$F(CH_2)_4OCHClCCl_3$
5-Fluoroamyl methyl ether	$F(CH_2)_5OCH_3$
6-Fluorohexyl methyl ether	$F(CH_2)_6OCH_3$
4,4′-Dichlorodibutyl ether (VI)	$Cl(CH_2)_4O(CH_2)_4Cl$

ω-FLUOROALKYL ETHERS

LD_{50} *(mice, I.P.) mg/kg*	*Conclusion*	*Boiling point*
15	toxic	56–57°
—	indefinite	92.5–93°/17 mm
60	indefinite	m.p. 49.5–50°
48	indefinite	94–95°/10 mm
43	indefinite	75–76°/14 mm
—	non-toxic	81°/21 mm
15–20	toxic	81°/22 mm
—	toxic	149°/13 mm
30–40	indefinite	132–133°/30 mm
10–20	toxic	103–104°/15 mm
70	indefinite	133–134°/12 mm
—	non-toxic	99–100°/12 mm
50	indefinite	164–167°
—	non-toxic	105–108°/13 mm
—	non-toxic	144–150°/13 mm
—	non-toxic	110–111°/14 mm
> 100	non-toxic	101–102°/12 mm
1.32	very toxic	100.5–101°/10 mm
1.5	very toxic	134.5–135°/10 mm
0.82	very toxic	73.5–74°/10 mm
6	toxic	124–125°/16 mm
90	non-toxic	127–128°
4.0	toxic	150–151°
> 100	non-toxic	128–130°/13 mm

to expectation, is approximately twice as toxic as the two mono-fluoro ethers. The intimate relationship of these compounds can clearly be seen by considering the LD_{50} figures in terms of mg *of fluorine*/kg:

	LD_{50} (mice, intraperitoneal)	
	mg of compound/kg	*mg of fluorine/kg*
4-Fluorobutanol	0.9	0.18
4-Fluoro-4′-chlorodibutyl ether (III)	1.32	0.14
4-Fluoro-4′-cyanodibutyl ether (IV)	1.5	0.16
4,4′-Difluorodibutyl ether (V)	0.82	0.18

Further confirmation of this mechanism is supplied by the observation[45] that 4-fluoro-4′-chlorodibutyl ether (III) caused a marked accumulation of citric acid in mice. This can best be explained by the following metabolic pathway, resulting in the formation of fluoroacetate, with its well-known sequelae:

$$F(CH_2)_4O(CH_2)_4Cl \rightarrow F(CH_2)_4OH \rightarrow F(CH_2)_3COOH \rightarrow FCH_2COOH$$

While the metabolic scission of methyl and ethyl *aromatic* ethers has long been recognized, little has been reported on the fate of *aliphatic* ethers *in vivo*. The toxicity results presented here provide circumstantial evidence for the rupture of ω-fluoroalkyl ethers in the animal body. It is probably not unreasonable to extend these conclusions to unfluorinated analogues by inferring the existence of a biochemical mechanism capable of promoting the rupture of simple aliphatic ethers.

Buckle and Saunders[10] have explained some of these results by an alternative process involving β-oxidation. Thus, in compounds of the type $FCH_2CH_2O(CH_2)_nCOOR$, if *n* is even, β-oxidation would yield 2-fluoroethyl hydrogen carbonate, FCH_2CH_2OCOOH, which would revert spontaneously to 2-fluoroethanol; in short the compound would be toxic. On the other hand, if *n* is odd, β-oxidation would yield 2-fluoroethoxyacetic acid, $FCH_2CH_2OCH_2COOH$, the ethyl ester (I) of which has been reported [10] to be non-toxic. Although the toxicity results are not entirely convincing, it is probable that both mechanisms (rupture and β-oxidation) may operate independently in some instances.

The insecticidal properties of various 2-fluoroethyl ethers, FCH_2CH_2OR ($R{=}CH_3$-, C_2H_5-, $CH_3CH_2CH_2CH_2$-, $BuOCH_2CH_2$-, CCl_3CHCl-, etc.)[54, 63, 64] and of 2-fluoro-2′-hydroxydiethyl ether, $FCH_2CH_2OCH_2CH_2OH$ and its derivatives[63, 64] have been examined. Some of the findings are described in Chapter 5 (p. 173).

ω-FLUORO COMPOUNDS CONTAINING NITROGEN OR SULPHUR

ω-Fluoronitriles, $F(CH_2)_nCN$

These compounds[50] may be prepared by dehydration of the appropriate amides, by reaction of an ω-fluoroalkyl halide with sodium cyanide, or by direct halogen exchange using an ω-bromonitrile:

$$\begin{array}{lll} (a) & F(CH_2)_nCONH_2 & \rightarrow F(CH_2)_nCN + H_2O \\ (b) & F(CH_2)_nX + NaCN & \rightarrow F(CH_2)_nCN + NaX \\ (c) & Br(CH_2)_nCN + KF & \rightarrow F(CH_2)_nCN + KBr \end{array}$$

Fluoroacetonitrile, the lowest member, has been reported to be non-toxic[9, 18], a fact implying that the nitrile is not hydrolyzed *in vivo* to the toxic fluoroacetic acid:

$$FCH_2CN \not\rightarrow FCH_2COOH$$

This lack of toxicity conforms to the reported metabolic breakdown of nitriles to the next lower acid and hydrogen cyanide[30]; the latter probably contributes to the overall effect, although the small quantity generated would slowly be converted to the non-toxic thiocyanic acid, HCNS by the natural detoxication mechanism.

$$RCH_2CN \rightarrow RCOOH + HCN$$

Fluoroacetonitrile would thus be expected to form the relatively innocuous fragments derived from the hypothetical fluoroformic acid. A simple means of confirming this breakdown mechanism

TABLE XXIII

Compound	*Formula*
Fluoroacetonitrile	FCH_2CN
3-Fluoropropionitrile	$F(CH_2)_2CN$
4-Fluorobutyronitrile	$F(CH_2)_3CN$
5-Fluorovaleronitrile	$F(CH_2)_4CN$
6-Fluorohexanonitrile	$F(CH_2)_5CN$
7-Fluoroheptanonitrile	$F(CH_2)_6CN$
8-Fluoro-octanonitrile	$F(CH_2)_7CN$
12-Fluorododecanonitrile	$F(CH_2)_{11}CN$

TABLE XXIV

Compound	*Formula*
3-Fluoro-1-nitropropane	$F(CH_2)_3NO_2$
4-Fluoro-1-nitrobutane	$F(CH_2)_4NO_2$
5-Fluoro-1-nitropentane	$F(CH_2)_5NO_2$
6-Fluoro-1-nitrohexane	$F(CH_2)_6NO_2$

was to examine the toxicological properties of the higher ω-fluoronitriles. It would naturally be expected that the ω-fluoronitriles containing an *odd** number of carbon atoms would be toxic, and those containing an *even* number, non-toxic.

The results shown in Table XXIII confirm this prediction: with the exception of fluoroacetonitrile and 4-fluorobutyronitrile, which are more toxic than expected, the toxicity of each ω-fluoronitrile is similar to that of the ω-fluorocarboxylic acid containing one less carbon atom. Moreover, it has been shown[45] that 7-fluoroheptanonitrile caused a large accumulation of citric acid in mice due to

* The terms 'odd' and 'even' refer to the *total* number of carbon atoms in the chain.

ω-FLUORONITRILES

LD_{50} *(mice, I.P.) mg/kg*	*Conclusion*	*Boiling point*
25	indefinite	79–80°
10	toxic	44–45°/12 mm
16	toxic	58.5–59°/16 mm
1.0	very toxic	71.5–72°/14 mm
50	indefinite	83.5–84°/11 mm
2.7	very toxic	97–98°/12 mm
> 100	non-toxic	117–118°/10 mm
80	non-toxic	114–115°/0.9 mm

ω-FLUORO-ω′-NITROALKANES

LD_{50} *(mice, I.P.) mg/kg*	*Conclusion*	*Boiling point*
92	non-toxic	69.5–70°/19 mm
11	toxic	78–79°/11 mm
90	non-toxic	98–99°/12 mm
12.5	toxic	106–106.5°/9 mm

formation of fluoroacetate; this can best be explained in terms of the C-CN rupture followed by β-oxidation:

$$F(CH_2)_6CN \rightarrow F(CH_2)_5COOH\ (+\ HCN) \rightarrow FCH_2COOH$$

In short, the toxicity results have provided independent verification of the mode of breakdown of aliphatic nitriles *in vivo*.

ω-Fluoro-ω′-nitroalkanes, $F(CH_2)_nNO_2$

The ω-fluoro-ω′-nitroalkanes[50] were prepared by the classical method of Victor Meyer:

$$F(CH_2)_nX + AgNO_2 \rightarrow F(CH_2)_nNO_2 + AgX \qquad (X = Br\ or\ I)$$

It has been shown[32, 65] that when nitroethane is administered intravenously to rabbits, the main metabolic products are acetal-

References p. 160

dehyde and nitrite. Assuming that the ω-fluoro-ω′-nitroalkanes follow the same metabolic pathway, the expected products would be the corresponding aldehydes. Since the ω-fluoroaldehydes are readily oxidized *in vivo* to ω-fluorocarboxylic acids (p. 132), of which only the even members are toxic, it seemed reasonable to predict that the even ω-fluoro-ω′-nitroalkanes would be more toxic than the odd members. The toxicity results shown in Table XXIV confirm this argument. Moreover, it has been shown[45] that 6-fluoro-1-nitrohexane results in the accumulation of citric acid in mice, due to formation of fluoroacetate:

$$F(CH_2)_6NO_2 \rightarrow F(CH_2)_5CHO \rightarrow F(CH_2)_5COOH \rightarrow FCH_2COOH$$

Hence all available evidence is in line with the biochemical breakdown of aliphatic nitrocompounds by the reaction

$$RCH_2NO_2 + O \rightarrow RCHO + HNO_2$$

ω-Fluoroalkylamines, $F(CH_2)_nNH_2$

2-Fluoroethylamine has been prepared[12, 39, 68] but no mention has been made of its toxicity. The higher members[53] were prepared by reduction of ω-fluoronitriles or ω-fluoro-ω′-nitroalkanes:

$$\text{(a)} \quad FCH_2)_nCN \rightarrow F(CH_2)_nCH_2NH_2$$
$$\text{(b)} \quad FCH_2)_nNO_2 \rightarrow F(CH_2)_nNH_2$$

In one instance, the acetyl derivative of an ω-fluoroalkylamine has been obtained[53] by the reaction of an ω-fluoroalkyl isocyanate *(q.v.)* with methylmagnesium chloride:

$$F(CH_2)_4NCO \xrightarrow{CH_3MgCl} F(CH_2)_4NHCOCH_3$$

Amine oxidase is a group-specific enzyme that occurs in small concentrations in many animal tissues. It catalyzes the oxidation of primary amines by the first of the reactions:

$$\begin{aligned} RCH_2NH_2 + O_2 &\rightarrow RCH = NH + H_2O_2 \\ RCH{=}NH + H_2O &\rightarrow RCHO + NH_3 \\ RCHO + O &\rightarrow RCOOH \end{aligned}$$

It is probable that the second reaction is not catalyzed but spontaneous, while the third reaction is a common enzymic oxidation. If this mechanism be applied to ω-fluoroalkylamines, it would be expected that the members containing an even number of carbon atoms would be toxic due to formation of the corresponding ω-fluorocarboxylic acids. Confirmation of this is supplied by the figures listed in Table XXV, thus providing independent verification of the above-mentioned metabolism of aliphatic amines. 4-Fluorobutylamine was too unstable for toxicity studies.

Parker and Walker[45] have carried out important investigations with the ω-fluoroalkylamines. Firstly, they showed that 6-fluorohexylamine produced a large accumulation of citric acid in mice and that 7-fluoroheptylamine produced none; this therefore provides confirmation that the toxic amines are degraded to the corresponding acids which in turn are β-oxidized to fluoroacetate. Secondly, they have noted that the amines were nearly as toxic when placed on the skin as when injected, in contrast with some other fluorine compounds examined; for example, the following approximate percutaneous toxicities have been determined using 6-fluorohexylamine[74]: rabbit, 0.25 mg/kg; rat, 1.5 mg/kg; cat, 0.9 mg/kg; and guinea pig, 1.4 mg/kg. Hence considerable caution must be observed in handling the toxic fluoroamines. Thirdly, it has been shown that the fluoroamines (in common with various unsubstituted alkylamines[74]) cause a characteristic red patch on the skin when applied percutaneously; this can be used as a subsidiary diagnostic aid.

If indeed the metabolism of the ω-fluoroalkylamines occurs as described above, it would be expected that 6-fluorohexylamine, 6-fluorohexanoic acid and fluoroacetic acid would all behave in an identical manner. Parker and Walker[45] have studied this problem by determining the citric acid levels in kidney and brain at various times after the administration of these three compounds; the amount of fluoroacetate used was approximately ten times greater than that of the other two compounds on a molar basis. Under these conditions the three compounds resulted in a very similar increase of citric acid with time, thus indicating that 6-fluoro-

References p. 160

TABLE XXV

Compound	*Formula*
3-Fluoropropylamine	$F(CH_2)_3NH_2$
4-Fluorobutylamine	$F(CH_2)_4NH_2$
N-4-Fluorobutylacetamide	$F(CH_2)_4NHAc$
5-Fluoroamylamine	$F(CH_2)_5NH_2$
6-Fluorohexylamine	$F(CH_2)_6NH_2$
7-Fluoroheptylamine	$F(CH_2)_7NH_2$
8-Fluoro-octylamine	$F(CH_2)_8NH_2$

hexylamine does not behave in a manner which is unique or different from that of fluoroacetate. However the smaller dose of 6-fluorohexylamine required to produce the same effect suggests that the fluoroamine is either excreted less rapidly or is metabolized to the toxic end-product (fluoroacetyl-coenzyme A) more efficiently than fluoroacetate.

One final piece of evidence in favour of the similarity of action of fluoroacetate and 6-fluorohexylamine was obtained by injection of these compounds into the carotid artery of a conscious dog. The symptoms in each case were the same, *i.e.* generalized convulsions after the usual latent period, followed by death from respiratory failure.

ω-Fluoro-α-aminoacids, $F(CH_2)_nCH(NH_2)COOH$

In a very recent report[58], Raasch has described two adjacent members of the ω-fluoro-α-aminoacid series, $F(CH_2)_nCH(NH_2)$-COOH. 5-Fluoronorvaline ($n = 3$) and 6-fluoronorleucine ($n = 4$) were prepared from the appropriate ω-fluoroalkyl bromides by treatment with diethyl sodioacetamidomalonate followed by hydrolysis using hydrofluoric acid:

$$F(CH_2)_nBr + \underset{\substack{|\\ NHAc}}{NaC(COOEt)_2} \rightarrow \underset{\substack{|\\ NHAc}}{F(CH_2)_nC(COOEt)_2} \rightarrow \underset{\substack{|\\ NH_2}}{F(CH_2)_nCHCOOH}$$

ω-FLUOROALKYLAMINES

LD_{50} *(mice, I.P.) mg/kg*	*Conclusion*	*Boiling point*
46	indefinite	88.5–89°
—	—	35–35.5°/65 mm
16.5	toxic	148–149°/13 mm
50	indefinite	61–61.5°/40 mm
0.9	very toxic	54–55°/13 mm
50	indefinite	67.5–68°/11 mm
0.76	very toxic	93–94°/15 mm

Toxicity determinations on mice by intraperitoneal injection indicated that 5-fluoronorvaline (m.p. 190°) was very toxic (LD_{50} 1.08 mg/kg) whereas 6-fluoronorleucine (m.p. 244°) was non-toxic ($LD_{50} > 215$ mg/kg). These clear-cut results may be explained by a degradative mechanism involving oxidative deamination, oxidative decarboxylation and β-oxidation:

$$F(CH_2)_3CH(NH_2)COOH \rightarrow F(CH_2)_3COCOOH \rightarrow F(CH_2)_3COOH \rightarrow FCH_2COOH$$

Such a mechanism, which postulates the intermediate formation of an α-ketoacid, is not necessarily invalidated by the low toxicity of fluoropyruvic acid, since this apparently behaves abnormally (p. 102); however, the preparation and examination of the ω-fluoro-α-ketoacids corresponding to the two fluoroaminoacids would certainly be desirable as a means of verifying the above degradative steps. The results as they stand provide yet another interesting example of the ω-fluorine atom acting as a 'tag' in the elucidation of intermediary metabolism.

ω-Fluoroalkyl thiocyanates, $F(CH_2)_nSCN$

The members examined[25, 59, 61] were prepared readily from the corresponding ω-fluoroalkyl halides or *p*-toluenesulphonates by reaction with potassium thiocyanate:

(a) $F(CH_2)_nX + KSCN \rightarrow F(CH_2)_nSCN + KX$

(b) $F(CH_2)_nOSO_2C_6H_4CH_3 + KSCN \rightarrow F(CH_2)_nSCN + CH_3C_6H_4SO_2OK$

Prior to this work, no mention had been made in the literature about the metabolic fate of aliphatic thiocyanates. It was considered likely, however, that reductive scission would occur, forming hydrogen cyanide and the corresponding mercaptans, and that the latter would then behave as described in the next section *(q.v.)*:

$$F(CH_2)_nSCN + 2H \rightarrow F(CH_2)_nSH + HCN$$

The alternation in toxicity of the ω-fluoroalkyl thiocyanates shown in Table XXVI is in accord with this suggestion, and thus provides evidence for the conversion *in vivo* of the thiocyanate grouping to mercaptan. That certain of the members, notably 3-fluoropropyl thiocyanate, are more toxic than the corresponding mercaptans may be associated with the concomitant formation of hydrogen cyanide. That 6-fluorohexyl thiocyanate ultimately forms fluoroacetate is shown[45] by the accumulation of citric acid following its administration to mice. This may be represented by the tentative scheme:

$$F(CH_2)_6SCN \rightarrow F(CH_2)_6SH \rightarrow F(CH_2)_6OH \rightarrow F(CH_2)_5COOH \rightarrow FCH_2COOH$$

TABLE XXVI

Compound	*Formula*
2-Fluoroethyl thiocyanate	$F(CH_2)_2SCN$
3-Fluoropropyl thiocyanate	$F(CH_2)_3SCN$
4-Fluorobutyl thiocyanate	$F(CH_2)_4SCN$
5-Fluoroamyl thiocyanate	$F(CH_2)_5SCN$
6-Fluorohexyl thiocyanate	$F(CH_2)_6SCN$
cf. n-Amyl thiocyanate	$CH_3(CH_2)_4SCN$

ω-Fluoroalkyl mercaptans, $F(CH_2)_nSH$

These mercaptans[25] were most conveniently prepared by reduction of the appropriate ω-fluoroalkyl thiocyanates by lithium aluminium hydride:

$$F(CH_2)_nSCN \xrightarrow{LiAlH_4} F(CH_2)_nSH + HCN$$

The acetates may be synthesized directly[25] by treatment of an ω-fluoroalkyl *p*-toluenesulphonate with potassium thiolacetate:

$$F(CH_2)_nOSO_2C_6H_4CH_3 + KSCOCH_3 \rightarrow F(CH_2)_4SCOCH_3 + CH_3C_6H_4SO_2OK$$

The products all had the characteristic obnoxious odour associated with the unfluorinated compounds.

By reference to the well-known metabolic transthiolation reaction ($RSH \rightarrow ROH$) it was considered likely that the ω-fluoroalkyl mercaptans would exhibit the same alternation in toxicity as that of the ω-fluoroalcohols (p. 127). That the two- and four-carbon members were acetylated was probably unimportant because of the widespread occurrence of deacetylating enzymes. The toxicity results shown in Table XXVII conform to the expected pattern and thus provide *a priori* evidence for the transthiolation *in vivo* of simple aliphatic mercaptans. Citric acid accumulation[45] resulting

ω-FLUOROALKYL THIOCYANATES

LD_{50} (mice, I.P.) mg/kg	*Conclusion*	*Boiling point*
15	toxic	78–79°/20 mm
18	toxic	75–76°/9 mm
2.6	very toxic	97–98°/13 mm
30	indefinite	112–113°/11 mm
5.0	toxic	124–125°/11 mm
75	non-toxic	90–91°/16 mm

TABLE XXVII

Compound	*Formula*
2-Fluoroethyl thiolacetate	$F(CH_2)_2SAc$
2-Fluoroethyl xanthate	$F(CH_2)_2SCSOEt$
'Sesqui-fluoro-H'	$F(CH_2)_2S(CH_2)_2S(CH_2)_2F$
3-Fluoropropyl mercaptan	$F(CH_2)_3SH$
4-Fluorobutyl thiolacetate	$F(CH_2)_4SAc$
5-Fluoroamyl mercaptan	$F(CH_2)_5SH$
6-Fluorohexyl mercaptan	$F(CH_2)_6SH$

from the administration of 6-fluorohexyl mercaptan once again confirms the formation of fluoroacetate.

$$F(CH_2)_6SH \rightarrow F(CH_2)_6OH \rightarrow F(CH_2)_5COOH \rightarrow FCH_2COOH$$

The fact that 'sesqui-fluoro-H', $F(CH_2)_2S(CH_2)_2S(CH_2)_2F$ was non-toxic and did not produce fluoroacetate-like symptoms suggests that at least some compounds containing the *thioether link* are stable in the animal body[60]. This compound is the fluorine analogue of sesqui-H, $Cl(CH_2)_2S(CH_2)_2S(CH_2)_2Cl$, a very potent member of the mustard group of blister gases; the absence of any vesicant activity adds emphasis to the fact that the action of the mustards is dependent upon reactive halogens.

ω-Fluoroalkanesulphonyl chlorides and fluorides, $F(CH_2)_nSO_2X$

The best method for preparing the ω-fluoroalkanesulphonyl chlorides involves the reaction of ω-fluoroalkyl thiocyanates with chlorine water[33, 59, 61]:

$$F(CH_2)_nSCN \xrightarrow{Cl_2\ aq.} F(CH_2)_nSO_2Cl$$

These in turn may be converted into the ω-fluoroalkanesulphonyl fluorides by treatment with aqueous potassium bifluoride[33]:

$$F(CH_2)_nSO_2Cl + KHF_2 \rightarrow F(CH_2)_nSO_2F + KCl + HF$$

ω-FLUOROALKYL MERCAPTANS AND DERIVATIVES

LD_{50} *(mice, I.P.)* *mg/kg*	*Conclusion*	*Boiling point*
56	indefinite	50–51°/20 mm
50	indefinite	208–210°
> 100	non-toxic	139°/17 mm
> 100	non-toxic	100–101°
1.8	very toxic	76.5–78°/13 mm
> 100	non-toxic	56°/17 mm
1.25	very toxic	68–69°/16 mm

The ω-fluoroalkanesulphonic acids, $F(CH_2)_nSO_2OH$ may be considered as the sulphonic acid analogues of the ω-fluorocarboxylic acids, $F(CH_2)_nCOOH$, and the two classes thus bear a superficial resemblance to one another. It was to find out if the two classes had similar toxicological properties that the ω-fluoroalkanesulphonyl derivatives were prepared and tested[33]. It has been reported that *n*-alkanesulphonic acids are non-toxic[26] and are excreted unchanged by the dog[14]; hence no great activity was anticipated for the ω-fluoro derivatives.

From an examination of the toxicity values presented in Table XXVIII, it can be seen that the sulphonyl chlorides exhibit a definite alternation in toxicity. However, this obviously cannot result from a β-oxidation mechanism analogous to that of the ω-fluorocarboxylic acids, since it is the members containing an *even* number of carbon atoms that are the more toxic:

$$\textit{e.g.}\quad \underset{\text{toxic}}{F(CH_2)_4\overset{\beta}{C}H_2\overset{\alpha}{C}H_2SO_2Cl} \not\longrightarrow \underset{\text{non-toxic}}{F(CH_2)_4COOH} + CH_3SO_2Cl$$

It seems more satisfactory therefore to explain the toxicity pattern in terms of the rupture of the carbon-sulphur bond (by some

TABLE XXVIII

Compound	*Formula*
2-Fluoroethanesulphonyl chloride	$F(CH_2)_2SO_2Cl$
2-Fluoroethanesulphonyl fluoride	$F(CH_2)_2SO_2F$
3-Fluoropropanesulphonyl chloride	$F(CH_2)_3SO_2Cl$
3-Fluoropropanesulphonyl fluoride	$F(CH_2)_3SO_2F$
4-Fluorobutanesulphonyl chloride	$F(CH_2)_4SO_2Cl$
4-Fluorobutanesulphonyl fluoride	$F(CH_2)_4SO_2F$
5-Fluoropentanesulphonyl chloride	$F(CH_2)_5SO_2Cl$
5-Fluoropentanesulphonyl fluoride	$F(CH_2)_5SO_2F$
6-Fluorohexanesulphonyl chloride	$F(CH_2)_6SO_2Cl$
6-Fluorohexanesulphonyl fluoride	$F(CH_2)_6SO_2F$
cf. *n*-Butanesulphonyl chloride	$CH_3(CH_2)_3SO_2Cl$
cf. *n*-Butanesulphonyl fluoride	$CH_3(CH_2)_3SO_2F$

mechanism as yet unknown), and of subsequent β-oxidation of the resultant ω-fluorocarboxylic acid.

$$\underset{\text{toxic}}{F(CH_2)_5CH_2SO_2Cl} \rightarrow \underset{\text{toxic}}{F(CH_2)_5COOH} \rightarrow FCH_2COOH$$

Once again, citric acid studies[45] confirm the ultimate formation of fluoroacetate from 6-fluorohexanesulphonyl chloride, presumably by the mechanism outlined above.

The correlation between structure and toxicity is less apparent in the case of the sulphonyl fluorides (Table XXVIII); such activity as they possess may be associated with cholinesterase inhibition[35].

Miscellaneous ω-fluoro compounds containing nitrogen

The ω-fluoro compounds[43] listed in Table XXIX comprise three isocyanates, two isothiocyanates and a symmetrical urea. The ω-fluoroalkyl isocyanates were prepared by the Curtius rearrangement:

$$F(CH_2)_nCOCl \xrightarrow{NaN_3} F(CH_2)_nCON_3 \rightarrow F(CH_2)_nNCO + N_2$$

ω-FLUOROALKANESULPHONYL CHLORIDES AND FLUORIDES

LD_{50} (mice, I.P.) mg/kg	*Conclusion*	*Boiling point*
19.5	toxic	82–83°/14 mm
8.8	toxic	62–63°/14 mm
64	indefinite	95.5–96°/12 mm
84	non-toxic	74–75°/15 mm
18	toxic	117–118°/13 mm
10	toxic	89–90°/12 mm
> 100	non-toxic	134–135°/13 mm
88	non-toxic	106–107°/12 mm
9	toxic	150–151°/15 mm
45	indefinite	128–129°/16 mm
> 100	non-toxic	83–86°/11 mm
70	indefinite	60–61°/17 mm

The isothiocyanates were obtained from the appropriate amines by the well-known reactions:

$$F(CH_2)_nNH_2 \xrightarrow[NaOH]{CS_2} F(CH_2)_nNHCSSNa \xrightarrow{ClCOOEt}$$

$$[F(CH_2)_nNHCSSCOOEt] \rightarrow F(CH_2)_nNCS + COS + EtOH$$

N,N′-Bis-4-fluorobutylurea was prepared from 4-fluorobutyl isocyanate by reaction with water.

$$2F(CH_2)_4NCO + H_2O \rightarrow F(CH_2)_4NHCONH(CH_2)_4F + CO_2$$

The ω-fluoroalkyl isocyanates are powerful lacrimators and are unstable in moist air. When stored under nitrogen in sealed ampoules, they exist as stable, colourless liquids. The paucity of members precludes much conclusive discussion in regard to their metabolism. It is interesting however to note the similarity in toxicity of 4-fluorobutyl isocyanate and N,N′-bis-4-fluorobutylurea, and that the

References p. 160

TABLE XXIX

Compound	*Formula*
2-Fluoroethyl isocyanate	$F(CH_2)_2NCO$
3-Fluoropropyl isocyanate	$F(CH_2)_3NCO$
4-Fluorobutyl isocyanate	$F(CH_2)_4NCO$
5-Fluoroamyl isothiocyanate	$F(CH_2)_5NCS$
6-Fluorohexyl isothiocyanate	$F(CH_2)_6NCS$
N,N′-Bis-4-fluorobutylurea	$[F(CH_2)_4NH]_2CO$

former is readily converted to the latter simply by treatment with water. These two observations lead to the suggestion that isocyanates may be initially converted *in vivo* to the corresponding symmetrical ureas before undergoing further change. That N,N′-bis-4-fluorobutylurea ultimately forms fluoroacetate is indicated by citric acid studies[45]. It is possible therefore that isocyanates may be metabolized by some such series of reactions:

$$F(CH_2)_4NCO \rightarrow F(CH_2)_4NHCONH(CH_2)_4F \rightarrow F(CH_2)_4NH_2 \rightarrow F(CH_2)_3CHO \rightarrow F(CH_2)_3COOH \rightarrow FCH_2COOH$$

Little can be said about the isothiocyanates. Parker and Walker have shown[45] that 6-fluorohexyl isothiocyanate results in citric acid accumulation in mice, but that 5-fluoroamyl isothiocyanate results in none. Consequently the isothiocyanate grouping is undoubtedly being removed metabolically (possibly by a mechanism akin to that suggested for the isocyanates).

DISCUSSION

The information presented in this chapter is both fragmentary and tentative. Obviously other homologous series containing the ω-fluorine atom could be found that would produce an alternation of toxicities similar to those described. Nevertheless, the series ω-FLUORO NITROGEN COMPOUNDS

LD_{50} *(mice, I.P.) mg/kg*	*Conclusion*	*Boiling point*
16.5	toxic	100–101°
10–20	toxic	126°
4.7	toxic	72–72.5°/42 mm
67	indefinite	104.5–105°/9 mm
11.2	toxic	116–116.5°/8 mm
4.4	toxic	m.p. 67–67.5°

already examined illustrate clearly the versatility of the technique for elucidating biochemical mechanisms and at the same time emphasize the striking ability of the mammalian organism to metabolize a remarkably wide range of foreign chemicals.

Much still remains to be done in preparing and studying these and related compounds. As remarked on numerous occasions, the biochemical conclusions require confirmation by independent methods: indeed, it is not unlikely that some may require extensive modification or even rejection. What may be considered as proved is the almost invariable characteristic toxicological pattern and the extremely hazardous nature of some of the members.

One general point should be mentioned here. It has become increasingly apparent in the study of ω-fluoro compounds that the lowest toxic member (*i.e.* the one forming fluoroacetic acid directly) is considerably less toxic than the higher toxic members in the same series[34]. This is borne out in the case of the various classes described in this chapter. There has been no report of work aimed specifically at the determination of the cause of this anomaly, which is essentially the same as that existing in the ω-fluorocarboxylic acid series (p. 94). The reason for it may well be connected with the relative ability of the members to form the 'active fluoroacetate' (fluoroacetyl-coenzyme A); this in turn may be dependent on the intramolecular electronic influence of the α-fluorine atom on the other functional groups, with resultant interference with metabolic breakdown or activation. In the opinion of the writer, this postulated electronic influence probably outweighs solubility and steric effects.

In attempting to formulate and apply rules for predicting the toxicity of new or unknown long-chain ω-fluoro compounds, it is necessary to know or to be able to guess the metabolic degradation of the functional 'head' in order to assess the eventual biological effect of the fluorinated 'tail'. Considering the general formula $F(CH_2)_nCH_2CH_2Z$ (where Z is any grouping), the toxicity is largely dependent on the fate of the $-CH_2CH_2Z$ grouping in the body. It is convenient to consider this under two classes:

(1) Z is removed ($-CH_2CH_2Z \rightarrow -CH_2COOH$)
(2) Z promotes removal of the adjacent methylene group ($-CH_2CH_2Z \rightarrow -COOH$)

In addition to effects of this sort, Z in some instances apparently facilitates chain rupture (for example in the case of the 1-fluoroalkanes and ω,ω'-difluoroalkanes), but this is probably of only secondary importance. If the fate of Z can be equated with that outlined in one of the above two classes, the following simple predictions emerge for any compound $F(CH_2)_nZ$:

Class of Z	*n in* $F(CH_2)_nZ$	*predicted effect*
1	odd	non-toxic
1	even	toxic
2	odd	toxic
2	even	non-toxic

Clearly, this rule is based on known degradations of different groups: a few representative examples (culled from a variety of sources, notably refs. 2 and 76) are therefore listed in Table XXX. It must be pointed out that some of these are tentative, but their inclusion is justified when taken in conjunction with the premise that it is better to consider a compound toxic that is not toxic, than *vice versa*.

TABLE XXX

METABOLIC DEGRADATION OF REPRESENTATIVE ALIPHATIC GROUPINGS (TENTATIVE)

Grouping Z	*Probable degradative route*
Class (1): Z alone is removed ($-CH_2CH_2Z \rightarrow -CH_2COOH$)	
$-C{\equiv}CH$	$-CH_2CH_2C{\equiv}CH \rightarrow -CH_2CH_2COCH_3 \rightarrow -CH_2COOH$
–Halogen	$-CH_2CH_2X \rightarrow -CH_2CH_2OH \rightarrow -CH_2COOH$
$-OH$	$-CH_2CH_2OH \rightarrow -CH_2COOH$
$-COCH_3$	$-CH_2CH_2COCH_3 \rightarrow -CH_2COOH$
$-OR$	$-CH_2CH_2OR \rightarrow -CH_2CH_2OH \rightarrow -CH_2COOH$
$-CN$	$-CH_2CH_2CN \rightarrow -CH_2COOH$
$-NO_2$	$-CH_2CH_2NO_2 \rightarrow -CH_2CHO \rightarrow -CH_2COOH$
$-NH_2$	$-CH_2CH_2NH_2 \rightarrow -CH_2CH{=}NH \rightarrow -CH_2CHO \rightarrow -CH_2COOH$
$-NCO$	$-CH_2CH_2NCO \rightarrow -CH_2CH_2NH_2 \rightarrow -CH_2COOH$
$-NCS$	$-CH_2CH_2NCS \rightarrow -CH_2CH_2NH_2 \rightarrow -CH_2COOH$
$-SH$	$-CH_2CH_2SH \rightarrow -CH_2CH_2OH \rightarrow -CH_2COOH$
$-SR$	$-CH_2CH_2SR \rightarrow -CH_2CH_2SH \rightarrow -CH_2CH_2OH \rightarrow -CH_2COOH$
$-SCN$	$-CH_2CH_2SCN \rightarrow -CH_2CH_2SH \rightarrow -CH_2CH_2OH \rightarrow -CH_2COOH$
$-SO_2X$	$-CH_2CH_2SO_2X \rightarrow -CH_2COOH$
$-SO_2OH$	$-CH_2CH_2SO_2OH \rightarrow -CH_2COOH$
$-SO_2R$	$-CH_2CH_2SO_2R \rightarrow -CH_2CH_2SO_2OH \rightarrow -CH_2COOH$
Class (2): Z promotes removal of adjacent CH_2 ($-CH_2CH_2Z \rightarrow -COOH$)	
$-CH_3$	$-CH_2CH_2CH_3 \rightarrow -CH_2CH_2COOH \rightarrow -COCH_2COOH \rightarrow -COOH$
$-COOH$	$-CH_2CH_2COOH \rightarrow -COCH_2COOH \rightarrow -COOH$
$-CHO$	$-CH_2CH_2CHO \rightarrow -CH_2CH_2COOH \rightarrow -COCH_2COOH \rightarrow -COOH$
$-CH(NH_2)COOH$	$-CH_2CH_2CH(NH_2)COOH \rightarrow -CH_2CH_2COCOOH \rightarrow -CH_2CH_2COOH \rightarrow -COCH_2COOH \rightarrow -COOH$
$-COCOOH$	$-CH_2CH_2COCOOH \rightarrow -CH_2CH_2COOH \rightarrow -COCH_2COOH \rightarrow -COOH$
$-CH(OH)COOH$	$-CH_2CH_2CH(OH)COOH \rightarrow -CH_2CH_2COCOOH \rightarrow -CH_2CH_2COOH \rightarrow -COCH_2COOH \rightarrow -COOH$
$-COCH_2COOH$	$-CH_2CH_2COCH_2COOH \rightarrow -CH_2CH_2COOH \rightarrow -COCH_2COOH \rightarrow -COOH$

In addition to the monofluoro compounds discussed above, various polyfluoro aliphatic compounds (notably polyfluoro-olefins) are toxic and pose a serious hazard to health. So far it has not been possible to formulate rules which embrace the known results (p. 124); until more data are available, the author would recommend that any aliphatic compound containing the $CF_2=$ grouping be handled with extreme caution.

REFERENCES

1. Anon. 'Teflon' TFE-fluorocarbon resins: safety precautions. *Bulletin No. X-59c, Polychemicals Department, E.I. du Pont de Nemours and Co. Inc.*, Wilmington, 98, Delaware.
2. BALDWIN, E. (1957) *Dynamic aspects of biochemistry*, Third Ed. Cambridge University Press.
3. BARTLETT, G. R. (1952) The mechanism of action of monofluoroethanol. *J. Pharmacol. Exptl. Therap., 106:* 464.
4. BERGMANN, E. D., and COHEN, S. (1958) Organic fluorine compounds. Part IX. The preparation of fluoroacetone and 1 : 3-difluoroacetone. *J. Chem. Soc., 1958:* 2259.
5. BERGMANN, E. D., and IKAN, R. (1957) The reaction of diazoketones with anhydrous hydrofluoric acid. *Chem. & Ind. (London), 1957:* 394.
6. BERGMANN, E. D., and SHAHAK, I. (1958) Transformation of toluene-*p*-sulphonates into fluorides. *Chem. & Ind. (London), 1958:* 157.
7. BERGMANN, F., and KALMUS, A. (1954) Synthesis and properties of ω-fluoroacetophenone. *J. Am. Chem. Soc., 76:* 4137.
8. BRICE, T. J., LAZERTE, J. D., HALS, L. J., and PEARLSON, W. H. (1953) The preparation and some properties of the C_4F_8 olefins. *J. Am. Chem. Soc., 75:* 2698.
9. BUCKLE, F. J., HEAP, R., and SAUNDERS, B. C. (1949) Toxic fluorine compounds containing the C-F link. Part III. Fluoroacetamide and related compounds. *J. Chem. Soc., 1949:* 912.
10. BUCKLE, F. J., and SAUNDERS, B. C. (1949) Toxic fluorine compounds containing the C-F link. Part VIII. ω-Fluorocarboxylic

acids and derivatives containing an oxygen atom as a chain member. *J. Chem. Soc., 1949:* 2774.

11. CHALLEN, P. J. R., SHERWOOD, R. J., and BEDFORD, J. (1955) 'Fluon' (polytetrafluoroethylene): a preliminary note on some clinical and environmental observations. *Brit. J. Ind. Med., 12:* 177.
12. CHILDS, A. F., GOLDSWORTHY, L. J., HARDING, G. F., KING, F. E., NINEHAM, A. W., NORRIS, W. L., PLANT, S. G. P., SELTON, B., and TOMPSETT, A. L. L. (1948) Amines containing 2-halogenoethyl groups. *J. Chem. Soc., 1948:* 2174.
13. EDGELL, W. F., and PARTS, L. (1955) Synthesis of alkyl and substituted alkyl fluorides from *p*-toluenesulfonic acid esters. The preparation of *p*-toluenesulfonic acid esters of lower alcohols. *J. Am. Chem. Soc., 77:* 4899.
14. FLASCHENTRÄGER, B., BERNHARD, K., LÖWENBERG, C., and SCHLÄPFER, M. (1934) Über einen neuartigen Abbau der aliphatischen Kette. *Hoppe-Seyler's Z. physiol. Chem., 225:* 157.
15. FRASER, R. R., MILLINGTON, J. E., and PATTISON, F. L. M. (1957) Toxic fluorine compounds. XV. Some ω-fluoro-β-ketoesters and ω-fluoroketones. *J. Am. Chem. Soc., 79:* 1959.
16. GITTER, S., BLANK, I., and BERGMANN, E. D. (1953) Studies on organic fluorine compounds. II. Toxicology of higher alkyl fluoroacetates. *Koninkl. Ned. Akad. Wetenschap. Proc. Ser. C, 56:* 427.
17. GREENBERG, L. A., and LESTER, D. (1950) Toxicity of the tetrachlorodifluoroethanes. *Arch. Indust. Hyg. Occupational Med., 2:* 345.
18. GRYSZKIEWICZ-TROCHIMOWSKI, E., SPORZYNSKI, A., and WNUK, J. (1947) Recherches sur les composés organiques fluorés dans la série aliphatique. II. Sur les dérivés des acides mono-, di- et trifluoroacétiques. *Rec. trav. chim., 66:* 419.
19. HAGEMEYER, D. R., and STUBBELINE, W. (1954) Physiological activity of fluorocarbon polymers. *Modern Plastics, 31, No. 12:* 136, 219.
20. HARRIS, D. K. (1951) Polymer-fume fever. *Lancet, 261:* 1008.
21. HOFFMANN, F. W. (1948) Preparation of aliphatic fluorides. *J. Am. Chem. Soc., 70:* 2596.
22. HOFFMANN, F. W. (1949) Aliphatic fluorides. I. ω,ω'-Difluoroalkanes. *J. Org. Chem., 14:* 105.
23. HOFFMANN, F. W. (1950) Aliphatic fluorides. II. 1-Halogeno-ω-fluoroalkanes. *J. Org. Chem., 15:* 425.

24. Howell, W. C., Cott, W. J., and Pattison, F. L. M. (1957) Organometallic reactions of ω-fluoroalkyl halides. II. Reactions of ω-fluoroalkylmagnesium chlorides. *J. Org. Chem., 22:* 255.
25. Howell, W. C., Millington, J. E., and Pattison, F. L. M. (1956) Toxic fluorine compounds. VII. ω-Fluoroalkyl thiocyanates and ω-fluoroalkyl mercaptans. *J. Am. Chem. Soc., 78:* 3843.
26. Kast, A. (1892) Zur Kenntniss der Sulfonalwirkung. *Naunyn-Schmiedeberg's Arch. exptl. Pathol. Pharmakol., 31:* 69.
27. Kitano, H., and Fukui, K. (1956) Fluoro ketones. *Kôgyô Kagaku Zasshi, 59:* 395.
28. Knapp, W. A. (1958) Personal communication, December 29, 1958.
29. Knunyants, I. L., Kisel', Ya. M., and Bykhovskaya, E. G. (1956) Reaction of hydrogen fluoride with diazo ketones. *Izvest. Akad. nauk S.S.S.R., Otdel. Khim. nauk, 1956:* 377.
30. Lang, S. (1894) Über die Umwandlung des Acetonitrils und seiner Homologen im Thierkörper. *Naunyn-Schmiedeberg's Arch. exptl. Pathol. Pharmakol., 34:* 247.
31. Lester, D., and Greenberg, L. A. (1950) Acute and chronic toxicity of some halogenated derivatives of methane and ethane. *Arch. Indust. Hyg. Occupational Med., 2:* 335.
32. Machle, W., Scott, E. W., and Treon, J. F. (1942) The metabolism of mononitroparaffins. I. Recovery of nitroethane from the animal organism. *J. Ind. Hyg. Toxicol., 24:* 5.
33. Millington, J. E., Brown, G. M., and Pattison, F. L. M. (1956) Toxic fluorine compounds. VIII. ω-Fluoroalkanesulfonyl chlorides and fluorides. *J. Am. Chem. Soc., 78:* 3846.
34. Millington, J. E., and Pattison, F. L. M. (1956) Toxic fluorine compounds. XII. Esters of ω-fluoroalcohols. *Can. J. Chem., 34:* 1532.
35. Myers, D. K., and Kemp, A. (1954) Inhibition of esterases by the fluorides of organic acids. *Nature, 173:* 33.
36. O'Brien, R. D., and Peters, R. A. (1958) The biochemistry of 2-deoxy-2-fluoro-DL-glyceraldehyde with a note on the toxicity of 2-deoxy-2-fluoroglycerol. *Biochem. Pharmacol., 1:* 3.
37. O'Brien, R. D., and Peters, R. A. (1958) The metabolism of DL-1-deoxy-1-fluoroglycerol. *Biochem. J., 70:* 188.
38. Olah, G., and Kuhn, S. (1956) Darstellung und Untersuchung

organischer Fluorverbindungen. XXI. Darstellung von Fluoracetaldehyd und aliphatischen Fluormethylketonen. *Chem. Ber.*, *89:* 864.

39. OLAH, G., and PAVLATH, A. (1955) Synthesis and investigation of organic fluorine compounds. XVII. Preparation of 2-fluoroethylamine. *Acta Chim. Acad. Sci. Hung.*, *7:* 461.
40. OLAH, G., PAVLATH, A., and NOSZKO, L. H. (1955) Synthesis and investigation of organic fluorine compounds. XIII. Derivatives of 2-fluoroethyl urethane. *Acta Chim. Acad. Sci. Hung.*, *7:* 443.
41. OLIVERIO, V. T., and SAWICKI, E. (1955) Some fluorine derivatives of urethan. *J. Org. Chem.*, *20:* 363.
42. OLIVERIO, V. T., and SAWICKI, E. (1955) New fluorinated urethans. *J. Org. Chem.*, *20:* 1733.
43. O'NEILL, G. J., and PATTISON, F. L. M. (1957) Toxic fluorine compounds. XIV. Some ω-fluoroalkyl nitrogen compounds. *J. Am. Chem. Soc.*, *79:* 1956.
44. PARKER, J. M., and WALKER, I. G. (1956) Personal communication, August 1, 1956.
45. PARKER, J. M., and WALKER, I. G. (1957) A toxicological and biochemical study of ω-fluoro compounds. *Can. J. Biochem. Physiol.*, *35:* 407.
46. PATTISON, F. L. M. (1953) Toxic fluorine compounds. I. *Nature*, *172:* 1139.
47. PATTISON, F. L. M. (1957) From war to peace: toxic aliphatic fluorine compounds. *Chem. in Can. 9*, *No. 8:* 27.
48. PATTISON, F. L. M., and BROWN, G. M. (1956) Organic nitrates as synthetic intermediates. Preparations of nitrates and some representative reactions. *Can. J. Chem.*, *34:* 879.
49. PATTISON, F. L. M., COTT, W. J., and HOWELL, W. C. (1957) Organometallic reactions of ω-fluoroalkyl halides. III. ω-Fluoroalkyllithium compounds. *J. Org. Chem.*, *22:* 464.
50. PATTISON, F. L. M., COTT, W. J., HOWELL, W. C., and WHITE, R. W. (1956) Toxic fluorine compounds. V. ω-Fluoronitriles and ω-fluoro-ω′-nitroalkanes. *J. Am. Chem. Soc.*, *78:* 3484.
51. PATTISON, F. L. M., and HOWELL, W. C. (1956) Toxic fluorine compounds. IV. ω-Fluoroalkyl halides. *J. Org. Chem.*, *21:* 748.
52. PATTISON, F. L. M., HOWELL, W. C., MCNAMARA, A. J., SCHNEIDER, J. C., and WALKER, J. F. (1956) Toxic fluorine compounds. III. ω-Fluoroalcohols. *J. Org. Chem.*, *21:* 739.

53. PATTISON, F. L. M., HOWELL, W. C., and WHITE, R. W. (1956) Toxic fluorine compounds. VI. ω-Fluoroalkylamines. *J. Am. Chem. Soc., 78:* 3487.
54. PATTISON, F. L. M., HOWELL, W. C., and WOOLFORD, R. G. (1957) Toxic fluorine compounds. XIII. ω-Fluoroalkyl ethers. *Can. J. Chem., 35:* 141.
55. PATTISON, F. L. M., and MILLINGTON, J. E. (1956) The preparation and some cleavage reactions of alkyl and substituted alkyl methanesulphonates. The synthesis of fluorides, iodides, and thiocyanates. *Can. J. Chem., 34:* 757.
56. PATTISON, F. L. M., and NORMAN, J. J. (1957) Toxic fluorine compounds. XVII. Some 1-fluoroalkanes, ω-fluoroalkenes and ω-fluoroalkynes. *J. Am. Chem. Soc., 79:* 2311.
57. PATTISON, F. L. M., STOTHERS, J. B., and WOOLFORD, R. G. (1956) Anodic syntheses involving ω-monohalocarboxylic acids. *J. Am. Chem. Soc., 78:* 2255.
58. RAASCH, M. S. (1958) 5-Fluoronorvaline and 6-fluoronorleucine. *J. Org. Chem., 23:* 1567.
59. SAUNDERS, B. C., and PATTISON, F. L. M. (1947) Unpublished work.
60. SAUNDERS, B. C., and STACEY, G. J. (1949) Toxic fluorine compounds containing the C-F link. Part IV. (a) 2-Fluoroethyl fluoroacetate and allied compounds. (b) 2,2′-Difluorodiethyl ethylene dithioglycol ether. *J. Chem. Soc., 1949:* 916.
61. SAUNDERS, B. C., STACEY, G. J., and WILDING, I. G. E. (1949) Toxic fluorine compounds containing the C-F link. Part II. 2-Fluoroethanol and its derivatives. *J. Chem. Soc., 1949:* 773.
62. SAWICKI, E., and RAY, F. E. (1953) Fluorourethan and derivatives. *J. Org. Chem., 18:* 1561.
63. SCHRADER, G. (1945) The development of new insecticides. Presented by MUMFORD, S. A. and PERREN, E. A. in *British Intelligence Objectives Sub-committee, Report No. 714.*
64. SCHRADER, G. *et al.* (1946) Developments in methods and materials for the control of plant pests and diseases in Germany. Reported by MARTIN, H., and SHAW, H. in *British Intelligence Objectives Sub-committee, Report No. 1095.*
65. SCOTT, E. W. (1942) The metabolism of mononitroparaffins. II. The metabolic products of nitroethane. *J. Ind. Hyg. Toxicol., 24:* 226.

66. STOKINGER, H. E. (1953) Teflon – a plastic with an inhalation hazard. *Occupational Health, 13:* 88.
67. TAYLOR, N. F., and KENT, P. W. (1956) The synthesis of 2-deoxy-2-fluorotetritols and 2-deoxy-2-fluoro-($\pm$)-glyceraldehyde. *J. Chem. Soc., 1956:* 2150.
68. TRAUBE, W., and PEISER, E. (1920) Über einige neue Umwandlungen des Äthylendiamins. *Ber., 53B:* 1501.
69. TREON, J. F., CAPPEL, J. W., CLEVELAND, F. P., LARSON, E. E., ATCHLEY, R. W., and DENHAM, R. T. (1955) The toxicity of the products formed by the thermal decomposition of certain organic substances. *Am. Ind. Hyg. Assoc. Quart., 16:* 187.
70. TREON, J. F., CLEVELAND, F. P., CAPPEL, J. W., and LARSON, E. E. (1954) The toxicity of certain polymers with particular reference to the products of their thermal decomposition. *Wright Air Development Center Tech. Rept., 54–301,* 60 pp.
71. TREON, J. F., CLEVELAND, F. P., LARSON, E. E., CAPPEL, J. W., SHAFFER, F., STEMMER, K. L., and DENHAM, R. T. (1954) The immediate toxicity of the thermal decomposition products of Teflon and Kel-F. *Report from the Kettering Laboratory, University of Cincinnati,* Cincinnati, Ohio, April 26, 1954.
72. TREON, J. F., STEMMER, K. L., LARSON, E. E., and CAPPEL, J. W. (1954) The toxicity of the products formed by the continuous thermal decomposition of Teflon. *Report from the Kettering Laboratory, University of Cincinnati,* Cincinnati, Ohio, April 30, 1954.
73. VOGEL, A. I., LEICESTER, J., and MACEY, W. A. T. (1956) *n*-Hexyl fluoride. In *Organic Syntheses,* vol. 36, John Wiley and Sons, Inc., New York, p. 40.
74. WALKER, I. G. (1958) Personal communication, September 30, 1958.
75. Walker, I. G., and PARKER, J. M. (1958) Further toxicological and biochemical studies on ω-fluoro compounds. *Can. J. Biochem. Physiol., 36:* 339.
76. WILLIAMS, R. T. (1947) *Detoxication mechanisms. The metabolism of drugs and allied organic compounds,* Chapman and Hall, London.
77. WILSHIRE, J. F. K., and PATTISON, F. L. M. (1956) Toxic fluorine compounds. XI. ω-Fluoroaldehydes. *J. Am. Chem. Soc., 78:* 4996.
78. ZAPP, J. A., LIMPEROS, G., and BRINKER, K. C. (1955) Toxicity of pyrolysis products of 'Teflon' tetrafluoroethylene resin. *Paper presented before American Industrial Hygiene Association,* April 28, 1955.

5

Potential uses and applications

Much of the work described in this chapter is tentative and subject to confirmation by independent means. The purpose of including such material in this monograph is merely to indicate trends and to offer clues for future researches. At the present stage of knowledge, only Compound 1080 (sodium fluoroacetate) can be considered as a proved industrial product, although the new inhalation anaesthetic Fluothane shows promise of being an outstanding addition to the anaesthetist's armamentarium.

Most industrial polyfluorinated compounds are innocuous *per se* but may give rise to toxic degradation products on excessive heating; such an effect is shown by the polymer Teflon (polytetrafluoroethylene) (see p. 125). However, since the original unpyrolyzed materials are non-toxic, it is inappropriate to discuss their preparation and uses in this monograph.

COMPOUND 1080 (SODIUM FLUOROACETATE) IN MAMMALIAN PEST CONTROL

The introduction in 1945 by the U.S. Fish and Wildlife Agency (Treichler and Ward) of sodium fluoroacetate (known in the United States as Compound 1080 or simply 'ten-eighty') as a rodenticide and general mammalian pest control agent[38] provided the stimulus for an extensive programme of study in regard to its toxicology, pharmacology and uses. Some of this work is reviewed in Chapter 2. Sodium fluoroacetate has been proved to be extremely dangerous to man[91] and to most household and farm animals[36, 90, 92], so its use in pest control is restricted to experienced exterminators. In addition to the inherent hazard of poisoned baits and water, the carcasses of deceased animals are very deadly to predators, since even a single dead mouse may contain enough poison to kill a full grown dog. It has been shown that dogs are killed by eating rats eight to ten weeks after they had been poisoned with sodium fluoroacetate[81]; again, dogs are killed by eating meat from a horse which had succumbed to methyl fluoroacetate[26]; whenever possible, carcasses must therefore be collected and destroyed. These facts, taken in conjunction with the general stability of the fluoroacetates under natural conditions, pose a serious problem for the ecologist, and stress the need for the utmost caution and foresight in the use of this type of poison.

Nevertheless, in skilled hands, sodium fluoroacetate is one of the most effective all-purpose rodenticides known[35]; it is fatal to rats regardless of their species (see Frontispiece). It is one of the few rodenticides that can be used successfully in water solution as well as in bait preparations; indeed there seems even to be a tendency for rats to prefer poisoned water to pure water, but this has not been definitely proved. It is not seriously subject to variations in potency, although pronounced deterioration of aqueous solutions occurs with time[12]. In short, the following features mark sodium fluoroacetate as being outstanding: high toxicity to all species of rodents tested, excellent acceptance, relatively quick action, absence of objectionable taste and odour, chemical stability, non-volatility, and no toxicity nor irritation on the skin of workers; moreover, rats apparently do not develop sufficient tolerance (p. 36) to the compound on ingestion of sublethal amounts to afford any appreciable protection, nor do they show significant reaction except after ingestion of lethal amounts; however, rats which survive poisoning occasionally develop an aversion to the compound which may become apparent during successive operations.

Jenkins and Koehler[35] have described the method of manufacture and distribution employed by the Monsanto Chemical Company, St. Louis, Missouri, the original American producers of Compound

1080*. In 1948, total annual production amounted to approx. 10,000 lbs. The essential process steps involve: (a) the reaction of purified ethyl chloroacetate with dry, powdered potassium fluoride in a stirred autoclave at 200° for about 11 hours; and (b) the conversion of the resultant ethyl fluoroacetate to sodium fluoroacetate in an agitated tank using a methanolic solution of sodium hydroxide. These authors have described the stringent safety precautions necessary in such a process. The commercial material is soluble in water, but relatively insoluble in organic solvents; it is hygroscopic and decomposes at about 200°. It is usually packaged in special metal containers with double friction-top closures and contains a nigrosine dye; the colouring serves as a means of ensuring even distribution in the preparation of the bait, as a satisfactory repellent for bird life, as a warning against using poisoned grain as a stock feed, and as a bait marker in field operations. Before an operator may purchase Compound 1080, he must satisfy the suppliers that he is reliable and that he carries a stipulated minimum public liability insurance policy. Owners of valuable animals who agree to the use of the poison must assume the obligation of preventing their animals from eating poisoned bait or deceased rodents.

This is not the place for an exhaustive survey of the voluminous technical literature relating to the use of sodium fluoroacetate and allied compounds in pest control, but it is worth recording the following random observations. Sodium fluoroacetate has been used successfully in the general extermination of rats, for example in towns[21] and ships[31–33]. This is of importance medically, since rats or their parasites play a primary rôle in the spread of bubonic plague, typhus fever and many less important illnesses, yet sodium fluoroacetate does not interfere with the guinea pig test for the diagnosis of plague[25]. It has been used to a lesser extent in the control of gophers and ground squirrels; and of coyotes[76] without endangering martens, weasels and minks[77]. It is being used on an increasing scale in Australia for rabbit extermination, now that

* Manufacture is now being carried out by the Pyrrole Chemical Corporation, Portsmouth, Ohio, and by the Tull Chemical Company, Oxford, Alabama.

resistance to myxomatosis has become widespread and general. Of the compounds related to sodium fluoroacetate, the following have been reported to be satisfactory as rodenticides: fluoroacetophenylhydrazide, $FCH_2CONHNHC_6H_5$[39, 40], fluoroacetamide, FCH_2CONH_2[11, 69, 70] and substituted fluoroacetamides, FCH_2CONHR and FCH_2CONRR'[8], 2-fluoroethanol, FCH_2CH_2OH[82, 83], urethanes of 2-fluoroethanol, $FCH_2CH_2OCONHR$ and $FCH_2CH_2OCONRR'$[82, 83], and the methylal of 2-fluoro-2'-hydroxydiethyl ether, $CH_2(OCH_2CH_2OCH_2CH_2F)_2$[82, 83]. The claim[72] that rats poisoned with fluoroacetamide die without convulsions has not been confirmed in other laboratories[68].

FLUORINE-CONTAINING ANAESTHETICS

In the general search for new non-explosive anaesthetics, a large number of fluorinated compounds have been examined in different laboratories. Most work has centred around fluorinated hydrocarbons and ethers[49, 75], the most promising compound in this early work being trifluoroethyl vinyl ether (Fluoromar), $CF_3CH_2OCH{=}CH_2$; however it was found to be explosive when mixed with oxygen in concentrations above 3%. The search was continued in the laboratories of Imperial Chemical Industries Ltd., culminating in the discovery of Fluothane. Suckling[85] has described the chemical and physical factors leading to its development, and Raventós[74] has outlined its anaesthetic and pharmacological properties.

Fluothane is 2-bromo-2-chloro-1,1,1-trifluoroethane, $CF_3\overset{\displaystyle Cl}{\overset{|}{\underset{\underset{\displaystyle Br}{|}}{CH}}}$.

It is a clear, colourless liquid (b.p. 50.2°) with a sweet, pleasant, non-irritant odour. When mixed in any concentration with oxygen or oxygen and nitrous oxide, it is non-explosive and non-inflammable. Although stable over soda-lime, it undergoes slow photochemical decomposition, which can be suppressed by the addition of a trace of thymol or by storage in brown bottles.

Fluothane operates by a biophysical mechanism in producing anaesthesia. Some of its favourable properties can be appreciated in the light of its chemical structure. The CF_3- grouping is well-known for its outstanding stability and for the fact that it reduces the reactivity of halogens on the adjacent carbon atom; hence Fluothane, which for these reasons is chemically inert, is not involved in metabolic reactions and consequently is non-toxic chemically; this is reflected in a fairly wide margin of safety. For the same reasons, it does not deteriorate appreciably on standing, when stabilized as described above. Finally, the low percentage of hydrogen in the molecule ensures that is is non-inflammable and non-explosive. It should be mentioned in passing that several other compounds containing the CF_3- grouping are known to be powerful anaesthetics, so adequate safeguards should be observed when working with compounds of this type.

Fluothane may be administered by all the conventional methods for inhalation anaesthetics. Most commonly it has been given using the principle of continuous flow of carrier gases (oxygen or oxygen plus nitrous oxide) through a suitable vaporizer. Its boiling point is sufficiently high to avoid the necessity of storage in a cylinder, yet sufficiently low to ensure easy vaporization and rapid excretion. It has been used with no serious post-operative illness in over 100,000 patients of all ages undergoing all types of surgical operation. The following are some of its characteristic features[5]: (a) Pleasant and rapid induction, making it especially valuable for children; (b) Smooth and easily reversible anaesthesia with a degree of relaxation adequate for most operations; (c) Suppression of salivary, bronchial and gastric secretions; (d) Absence of the shock syndrome during and immediately following surgery; (e) Very rapid and uneventful recovery, with relative freedom from nausea and vomiting.

It must be noted however that its potency and speed of action can rapidly lead to overdosage if it is mishandled; to date, at least two deaths have resulted from its use or misuse, with pathological changes similar to those produced by chloroform. Moreover, Krantz and colleagues[45] in recent work with mice, dogs, and monkeys, have found that its margin of safety is narrower than that reported by Raventós[74], and that its depressor response and depression of oxygen consumption appear to be serious disadvantages in its use as an anaesthetic.

From the above description, it will be clear that the pharmacological action of Fluothane is entirely different from that of the majority of compounds described in this monograph, in that it operates by a physical and not a chemical mechanism. Indeed, in common with most polyfluoro compounds it is non-toxic chemically, and its inclusion is justified solely on the basis of its being a prototype of potentially dangerous aliphatic fluorine compounds. In this same category comes Indoklon, an inhalant convulsant, which is described briefly in the next section.

INDOKLON, A FLUORINE-CONTAINING CONVULSANT

Indoklon (hexafluorodiethyl ether, $CF_3CH_2OCH_2CF_3$) in common with Fluothane (p. 169) is a relatively non-toxic polyfluoro compound but gives rise to a pronounced pharmacological response. It is appropriate that some account be given of its action on man to illustrate the unexpected biological effects which may be encountered in work with aliphatic fluorine compounds.

Krantz and colleagues, in their investigation of aliphatic fluorinated ethers as anaesthetics[49], observed that hexafluorodiethyl ether (Indoklon) elicited violent convulsions upon inhalation in many species of laboratory animals[46]. The compound is a colourless, volatile, non-inflammable, pleasant-smelling liquid of b.p. 64°. Because of many favourable properties, including rapid onset of seizures, ease of control of depth and duration of the seizures, and the apparently harmless nature of repeated exposures, it was suggested[46] that Indoklon might find use as an alternative to electroconvulsive therapy in the treatment of certain types of mental illness.

Clinical results have been reported recently[22, 44] for 80 patients randomly selected for convulsive therapy. To date, no untoward results have been observed. In the 1000 treatments administered, there was no failure in bringing about a *grand mal* convulsion. The seizure was similar to that produced by electric stimulation, except for a more gradual onset. Subjectively, many patients who had experienced this procedure and who had had electroconvulsive therapy in the past, while not feeling enthusiastic about either treatment procedure, seemed to feel less threatened by the inhalation technique. This may be due to the fact that the seizure is invariably preceded by unconsciousness.

The quantity of Indoklon vapour required to bring about a convulsion averages the equivalent of 0.5 c.c. of the liquid, with no increased sensitivity or tolerance noted with progressive treatments. An induction period of about 20 seconds is optimal, since longer periods create anxiety in the patients as they await unconsciousness; convulsions cannot be elicited in less than 10 seconds. The onset of the seizure is heralded by quivering and fluttering movements of the eyelids and mild myoclonic jerks. The beginning of the subsequent tonic phase of the convulsion is indicated by the opening of the jaws. This preliminary phase of activity may prevent the sudden muscular contractions and 'jackknifing' which are thought to produce the fractures associated with electroconvulsive therapy. The tonic phase usually lasts 15 to 30 seconds, and the subsequent clonic phase 15 to 35 seconds. Total recovery following the seizure ranges from 2 to 15 minutes. An overdose (5 c.c., *i.e.* a ten-fold increase) in one patient produced a comatose state, but was not fatal. The mechanism by which the convulsive seizure is brought about is unknown.

In psychiatric therapy, the use of Indoklon appears to be at least as effective as electric stimulation, with apparently less apprehension and less post-convulsive confusion. Thus in a group of 25 patients with mental disorders in which severe depression or schizophrenic reactions were the outstanding characteristics, 9 showed a sufficient degree of improvement to be released from hospital, 5 were greatly improved, 10 derived only transitory benefit, and one displayed no change during treatment[44].

CHEMICAL WARFARE AGENTS

During the last war, the fluoroacetates were examined as candidate chemical warfare agents[52, 60, 67, 78, 79] but, in common with all other war gases, were not used for this purpose. It has been pointed out earlier (p. 1) that the fluoroacetates are insidious poisons for which no antidote is known. Not unnaturally, the properties responsible for such action provide a recommendation for their use in warfare; some of these may be listed now. (a) They are highly toxic. (b) They operate with a delayed action, thus ensuring that a lethal quantity can be ingested before toxic symptoms develop. (c) They are inconspicuous in regard to taste, smell and total lack of immediate symptoms. (d) No medical treatment is known which is likely to save the life of the casualty after symptoms have developed. (e) They are sufficiently stable to survive most of the usual rather drastic methods of dispersion. (f) Their stability renders field detection and decontamination extremely difficult, and presents problems in regard to protection of personnel*. (g) Many of them have satisfactory physical constants for use as persistent or semi-persistent war gases; moreover, the volatility of the toxic fluoroacetate and ω-fluorocarboxylate esters may be varied at will by the choice of appropriate alcoholic moieties.

It is probable that the well-known ‘nerve gases’ and ‘blister gases’[4] would be more suitable for large-scale use in chemical warfare. The value of the fluoroacetates and related compounds might lie more in specialized undercover activities such as poisoning of vital water supplies. Indeed, a toxic fluoroacid was employed for this very purpose many years ago by the natives of Sierra Leone in their use of ratsbane for contaminating streams which supplied hostile villages (p. 86).

INSECTICIDES

While many different types of aliphatic fluorine compounds have activity as fumigant and contact insecticides[6, 8, 29, 30, 34, 54, 55, 62, 63, 66, 82, 83, 86], the most important developments are likely to occur in the field of *systemic insecticides*. These are substances which are absorbed by a plant and are subsequently translocated

* Most gas masks afford protection against aliphatic fluorine compounds.

References p. 187

throughout its system. The insecticide may be applied to the plant itself, to the roots, to the soil around the plant, and in some cases even to the seed of the plant. It is sometimes not the systemic insecticide *per se* that is the active material, but rather the metabolic product of the insecticide resulting from some enzymic process in the plant. The absorbed agents or the resulting metabolic products kill suctorial and leaf-eating insects without killing the plant. Schrader [82, 83, 87] in his pioneer work, noted that this type of insecticide is particularly valuable for combatting insects which attack plants in inaccessible sites, such as the roots; thus, for general purposes, they are often superior to contact insecticides which are applied topically by spraying.

Prior to the development of systemic insecticides, Schrader examined a wide range of fluorine compounds as contact and fumigant insecticides for use mainly as nicotine substitutes; the following were reported to be particularly effective: methanesulphonyl fluoride, CH_3SO_2F, thionyl fluoride, SOF_2, dimethylcarbamyl fluoride, $(CH_3)_2NCOF$, and dimethylamino-sulphuryl fluoride, $(CH_3)_2NSO_2F$. It is interesting to note that bis-2-fluoroethyl sulphite, $(FCH_2CH_2O)_2SO$ was found to be a potent insecticide for sucking insects yet showed no phytotoxicity; the structure and properties of this material seem to set the stage for the discovery of the first true systemic insecticides.

From paraformaldehyde, the appropriate fluoroalcohol and a trace of sulphuric acid, the two following methylals were easily produced:

$$CH_2\left\langle\begin{matrix}OCH_2CH_2F\\OCH_2CH_2F\end{matrix}\right. \qquad CH_2\left\langle\begin{matrix}OCH_2CH_2OCH_2CH_2F\\OCH_2CH_2OCH_2CH_2F\end{matrix}\right.$$

Shown originally to be typical contact insecticides, it was observed that they also functioned as 'chemotherapeutic agents for living plants'[82]; that is, they were readily absorbed by the plant with no ill effect, and thus conferred a sufficiently high toxicity to the roots, stems and leaves as to kill predatory insects, such as plant-feeding caterpillars and aphids, and even rabbits. The effect lasted for 3 to 4 weeks. With this discovery, 'a problem which has been worked on for decades nears its solution'[82]. The term 'systemic insecticide' was subsequently coined by Martin and Shaw[83].

The information thus acquired was turned at once to combatting a very serious infestation[82, 83]. The entire vine industry in Germany was suffering under an attack of *phylloxera*. The recognized treatment for this involved sprinkling all infected vines with carbon disulphide; thus all vines were destroyed and only after four years could the area be replanted. Such ruthless measures were necessary because *phylloxera* attacks the roots, and hence could not be eradicated by sprays. The loss to the vineyard owner was disastrous. However, trials with the second of the above methylals showed that *phylloxera* could be effectively controlled without necessitating the destruction of the vines; the harvested grapes were of course poisonous and could not be used, but the vines were preserved and could safely be used the following season. It was only the difficulty of preparing the intermediate 2-fluoroethanol in quantity that prevented large-scale testing of these materials. With the more recent discoveries of the excellent phosphorus-containing systemic insecticides, the above work becomes only of historical interest.

In 1950, David[13] reported that sodium fluoroacetate was very effective as a systemic insecticide; this observation was confirmed in later work with various different insects including *Aphis fabae*[14], *Pieris brassicae*[15] and *Phaedon cochleariae*[16]. Other workers have described the efficacy of fluoroacetates as insecticides using, for example, fleas[50], moths[88], mosquitos[19], and sweet-potato weevils[28]. A valuable property of certain phosphorus-containing insecticides is that of conferring protection on plants grown from seeds which had previously been soaked in a solution of the insecticide; this property is not possessed by sodium fluoroacetate[17]. In spite of some very attractive properties, the hazardous nature of sodium fluoroacetate precluded its use on a commercial scale. Since a variety of long-chain esters derived from fluoroacetic acid and from 2-fluoroethanol are related structurally to the compounds tested by Schrader and by David but are less toxic to mammals[62], they were examined and found to possess high activity as systemic insecticides[34, 62]; among the most effective was 2-fluoroethyl

References p. 187

laurate, $FCH_2CH_2OCO(CH_2)_{10}CH_3$[59], which had low phytotoxicity and relatively low mammalian toxicity, yet caused 88% mortality of aphids at only 10 p.p.m.

Rather more effective than these esters were some fluorinated ethers[63], most outstanding of which was 2-fluoro-1′, 2′, 2′, 2′-tetrachlorodiethyl ether, $FCH_2CH_2OCHClCCl_3$[61]. This molecule combines the toxiphoric 2-fluoroethyl group with the lipophilic $-CCl_3$ group associated with the well-known insecticide DDT; the ether linkage combining these two groups can be split *in vivo* (see p. 139) yet is apparently sufficiently stable to survive translocation in the plant. Preliminary testing indicates that this compound is active systemically against aphids at 1.3 p.p.m.[63], but is much less effective against most other insects and larvae[89] in regard both to systemic and to contact activity. Its mammalian toxicity is relatively low (see p. 140), and it did not damage the plants at the concentrations examined (0.1% and less).

Other compounds which have been reported to have insecticidal activity include fluoroacetamide, FCH_2CONH_2[6, 18, 66, 71], N-monosubstituted fluoroacetamides, FCH_2CONHR[8, 70], fluoroacetophenylhydrazide, $FCH_2CONHNHC_6H_5$[86], fluoroacetate esters, FCH_2COOR[66, 93], 2-butoxyethyl fluoroacetate, $FCH_2COOCH_2CH_2$-OBu[29], 2-ethylhexyl fluoroacetate, $FCH_2COOCH_2CHEt(CH_2)_3$-CH_3[30], and various derivatives of fluoroethanol[54, 55]. Of these, fluoroacetamide as a 1% aqueous solution has recently been marketed in Britain as a systemic insecticide (Tritox) for use with non-edible plants[71]. The formulation includes acetamide, which, it is claimed, minimizes danger of accidental poisoning by reducing the mammalian toxicity[72]; this is hardly surprising, however, since acetamide is known to afford protection against fluoroacetates in general (see p. 37).

In conclusion it should be repeated that all fluorinated aliphatic insecticides have appreciable mammalian toxicity (some more than others), and hence that their usefulness is likely to be limited to specialized applications, such as protecting non-edible crops (tobacco, etc.). For the same reason, extreme caution should be observed at all times when working with compounds of this type.

ORGANIC FLUORO COMPOUNDS AS A SOURCE OF FLUORIDE

The pharmacology of inorganic fluorides, and, more particularly, the highly controversial subject of fluoridation of drinking water, can have no place in a book devoted to organic compounds. However, mention must be made of a report by Ott, Piller and Schmidt[57] in which it is claimed that derivatives of 5-fluorovaleric acid function as a convenient source of fluoride for hardening the enamel of the teeth. Thus, in experiments with rats fed with a mixture of the two monoglycerides derived from this acid (in a dosage corresponding to 1 mg of fluorine per day), it was shown that the fluorine was rapidly absorbed and that the fluoride content of the incisors was increased. By this method, it is claimed that fluoride can be assimilated easily and without the troubles in the calcification of the teeth associated with inorganic fluorides. Hence it is suggested that ingestion of these derivatives may be of greater value in reducing dental caries than the commoner procedure of incorporating inorganic fluorides into urban water supplies.

Assuming that the organic samples were free from traces of contaminating fluoride, this observation can only be explained in terms of the rupture of a C-F bond *in vivo*. Such fission may possibly be aided by some as yet unrecognized micro-organism or enzyme system (see p. 16); alternatively, it may take place spontaneously after the occurrence of some preliminary degradative steps, for example:

$$F(CH_2)_4COOH \xrightarrow{\beta\text{-oxidation}} [FCOOH] \xrightarrow{\text{spontaneous}} HF + CO_2$$

Both these suggestion are speculative, and entirely unsupported by experimental evidence.

FLUOROACETATE AS AN ANTIDOTE TO LEAD POISONING

The citrate ion (Cit^{3-}) forms a stable complex ($PbCit^{-}$) with divalent lead under physiological conditions; thus, sodium citrate in clinical use has resulted in symptomatic relief in some cases of lead poisoning. However, large frequent doses were required to counteract the rapid disappearance of citrate by excretory and metabolic pathways; moreover, its distribution is primarily extracellular. Consequently, Fried, Rosenthal and Schubert[24] explored the idea of inducing an abnormally high *intracellular* concentration of citrate by the administration of a sublethal dose of fluoroacetate.

It was found that, by this technique, partial protection was afforded to rats acutely poisoned with lead nitrate. Of rats given an LD_{90} of lead nitrate, 53% survived when treated with sodium fluoroacetate, whereas only 10% survived in the control group. The LD_{50} of lead nitrate was thus increased from 58.2 mg/kg (as Pb) to 67.7 mg/kg in fluoroacetate-treated rats. It was found that kidney damage was one of the main causes of death; it was therefore suggested that the major protective effect of fluoroacetate resulted from the elevation of intracellular citric acid levels in the kidney.

Although only partial protection was achieved, the results demonstrate the feasibility of the general method, which involves an interference with metabolic processes *in vivo*, such that certain compounds, which are naturally present in the body and which possess chelating or complexing properties, are maintained at an abnormally high concentration within the cells and thus are utilized to modify metal toxicity. This use of fluoroacetate affords an interesting example of a sublethal dose of one toxic agent protecting against acute poisoning by another.

FLUOROACETATE AS A PROTECTING AGENT AGAINST IRRADIATION

The mortality and somatic damage caused to living organisms by X- or γ-radiation are greatly reduced by the previous administration of various compounds known as radio-protectors.

One of the most serious consequences of irradiation is thought to be the destruction of deoxyribonucleic acid by deoxyribonuclease. The fact that this enzyme requires magnesium ions for its activity led to the ingenious suggestion of Bacq, Fischer, Herve, Liébecq and Liébecq-Hutter[7] that agents capable of complexing magnesium ions would reduce the activity of the enzyme and might protect against irradiation. Large doses of sodium citrate were ineffective, probably due to failure to penetrate essential tissue barriers, but sublethal doses of sodium fluoroacetate, which induce high concentrations of citrate *in situ*, were found to afford significant protection[7]. For example, the administration of an LD_{50} dose of sodium fluoroacetate (4–5 mg/kg) to mice 4 to 5 hours before irradiation yielded the following results: after 7 days, 42% of the unprotected irradiated mice had died whereas all the preinjected mice were alive; after 14 days, the mortality figures for the two groups were 90% and 20%. Protection was afforded *only* when the fluoroacetate was administered a few hours before irradiation, corresponding to the time necessary for the citric acid accumulation to reach a maximum. This therefore supports (but does not prove) the above suggestion that fluoroacetate may exert its protective action by accumulating citrate, complexing the intracellular magnesium ions, affecting the activity of deoxyribonuclease and hence reducing the destruction of deoxyribonucleic acid. It is possible that fluoroacetate may also protect by reducing the sensitivity of the cells to X-rays through interference with their oxidative metabolism (many radio-protectors are in fact toxic agents). Further work will be necessary before the detailed mechanism is fully substantiated. It is of course very doubtful if the possible therapeutic value of this treatment justifies the risk of fluoroacetate poisoning, particularly when there are safer and more effective radio-protectors known.

FLUOROACETATE AS A SPECIFIC INDICATOR OF THE TRICARBOXYLIC ACID CYCLE

Peters has reviewed in detail the uses of fluoroacetate as a specific indicator of the tricarboxylic acid cycle[68]. If, after administration of fluoroacetate, there is a pronounced accumulation of citric acid (usually accompanied by a simultaneous decrease in oxygen uptake), then it may be concluded that the tricarboxylic acid cycle operates in the system under study. The symptoms associated with fluoroacetate poisoning in fact serve as a marker of the tricarboxylic acid cycle.

The method may be illustrated by three recent examples. (a) Strauss[84] has observed that mutants of *Neurospora crassa* requiring dicarboxylic acids for growth will oxidize acetate. That this oxidation proceeds through the agency of the tricarboxylic acid cycle was indicated by an accumulation of citric acid when fluoroacetate was added to the growing cultures. (b) Dietrich and Shapiro[20], in studying homogenates of seven different mouse neoplasms, observed that fluorocitrate in *all* instances resulted in a large accumulation of citric acid. However, of the seven, *only* in the adenocarcinoma 755 (grown in the C57BL mouse) did fluoroacetate cause citrate accumulation in any way comparable to that caused by fluorocitrate. The first observation indicates that all seven tumours possess aconitase activity (which is subsequently inhibited by the fluorocitrate with resultant accumulation of citric acid) and hence that they derive energy from the tricarboxylic acid cycle; on the other hand, the second observation indicates that, apart from the adenocarcinoma 755, mouse neoplastic tissue cannot effectively metabolize fluoroacetate, and hence that there is some deficiency in the mechanism responsible for the entry of C_2 fragments into the cycle. (c) Sherwood Jones, Maegraith and Gibson[37] have provided evidence for the operation of the tricarboxylic acid cycle in the rat reticulocyte; thus, it was shown that fluoroacetate inhibited pyruvate oxidation and caused an accumulation of citrate.

POTENTIAL VALUE OF ALIPHATIC MONOFLUORO COMPOUNDS AS DRUGS

No outstanding aliphatic fluorine-containing drug has yet been developed; and, from a consideration of the title of this monograph, it is clear that no mention can be made of such promising non-aliphatic medicinal products as the interesting and topical fluorinated steroids and heterocyclic compounds. It is the opinion of the writer however that it is only a question of time before discoveries are made in the aliphatic field along the lines illustrated in the following example[64].

One of the most important conclusions arising from the toxicological studies of aliphatic monofluoro compounds is that many organisms are unable to differentiate between compounds containing the methyl (CH_3-) and fluoromethyl (FCH_2-) groups respectively; reasons for this are given on p. 26. Compounds containing the latter group, once assimilated, may then disturb the delicate balance of metabolic processes.

Occurring in the lipid sheath of the tubercle bacillus are a variety of fatty acids, including the laevo-rotatory isomer of 10-methyloctadecanoic acid (tuberculostearic acid), $CH_3(CH_2)_7CHMe$-$(CH_2)_8COOH$. By applying the above observations to this acid, simple substitution of the terminal methyl group by the fluoromethyl group gives rise to ω-fluorotuberculostearic acid, $FCH_2(CH_2)_7CHMe(CH_2)_8COOH$. It was argued that this might readily be assimilated by the tubercle bacillus, and that the toxic action of the organically-bound fluorine atom might then result in its death.

The acid was prepared[64] by anodic coupling and tested against virulent *Mycobacterium tuberculosis* var. *hominis* (strain H37Rv) both *in vitro* and *in vivo*. In the former, it showed high activity, causing complete inhibition of growth at a concentration of 1.25 γ/ml. However it did not prolong the lives of tuberculous mice, and was therefore inactive in what must be considered the crucial test. Nevertheless, it is probably not fortuitous that of some thirty-five ω-fluoro compounds screened against the tubercle bacillus, the only one to show high activity *in vitro* was this acid; the other compounds had been selected at random, whereas this acid was devised and synthesized specifically for this purpose.

In summary, this approach to the development of new drugs involves: (a) the selection of an appropriate essential biochemical constituent of the pathogenetic organism to be attacked; and (b) the preparation of this constituent modified by the judicious introduc-

References p. 187

tion of a fluorine atom in the molecule. It is of course essential that the toxicity differential should favour the host.

Ancillary to this technique is the development of fluorinated antimetabolites; indeed, the compounds devised by the above procedure may be considered as fluorinated antimetabolites of the organism under attack. More generally, the presence of a fluorine atom in any essential biological compound tends to cause symptoms characteristic of a deficiency of the unfluorinated compound. For example, 3-fluorotyrosine competes with tyrosine in thyroxine formation and consequently has potent antithyroid activity[47, 48]. Chenoweth[12] has concluded that a fluorinated compound must have a structure capable of partially, but *not completely*, mimicking its natural analogue if it is to behave effectively as an antimetabolite. (Were there no physico-chemical differences at all, there would seem to be no reason to expect any pharmacological differences.) Because of the remarkable spatial similarity of the fluorine and hydrogen atoms, a fluorinated hormone may enter the reactive centre specific for the non-fluorinated hormone, but, for reasons associated with the physical properties of the fluorine atom, cannot readily be dislodged; the enzyme is thus blocked, and symptoms develop similar to those produced by a simple deficiency of the unfluorinated hormone. That fluorinated hormones or precursors result in deficiency symptoms can be used to advantage for countering glandular hyperactivity.

ELUCIDATION OF BIOCHEMICAL PROCESSES

The toxicity pattern of the series of ω-fluorocarboxylic acids, $F(CH_2)_nCOOH$ described in Chapter 3, was correlated with the β-oxidation theory of fatty acid metabolism (p. 90). It was pointed out that the corollary to this explanation furnished novel and independent verification of the β-oxidation theory. Thus, the ω-fluorocarboxylate series may be considered as the first ω-fluoro series to provide indirect evidence regarding a biochemical process. Some of the series described in Chapter 4 have been of value in confirming the known detoxication of a variety of functional groups, and other series have been used to *predict* the metabolism of compounds or groupings of which the mode of breakdown was not known. Examples of the latter included the 1-fluoroalkanes, ω, ω'-difluoroalkanes, ω-fluoroalkynes, ω-fluoroalkyl halides, ω-fluoroketones, ω-fluoroethers, and ω-fluoroalkyl thiocyanates.

The method is based merely on crude toxicity figures, the rationale of which resides in the characteristic toxicological behaviour of the ω-fluorocarboxylic acids. In the study of the metabolic fate of any grouping Z in an aliphatic compound, the first task is the preparation of a few adjacent members of the series $F(CH_2)_nZ$; thereafter, the examination of the toxicity pattern allows of the interpretation of the results. The ω-fluorinated 'tail' thus acts as a 'tag' for deducing the metabolism of the functional 'head'. The type of reasoning may be illustrated by a consideration of the nitrile grouping (-CN). It has been shown on p .144 that the nitriles containing an *odd* number of carbon atoms are toxic (*e.g.* $F(CH_2)_4$-CN) and those containing an *even* number are non-toxic (*e.g.* $F(CH_2)_5CN$). Listed below are three possible modes of degradation of the nitrile grouping which might occur to the organic chemist.

(a) *Hydrolysis:* $RCN \rightarrow RCOOH$
e.g. $F(CH_2)_4CN \rightarrow F(CH_2)_4COOH$
toxic — non-toxic

(b) *Reduction:* $RCN \rightarrow RCH_2NH_2$
e.g. $F(CH_2)_4CN \rightarrow F(CH_2)_5NH_2$
toxic — non-toxic

(c) *Cleavage of C-CN:* $RCH_2CN \rightarrow RCOOH + HCN$
e.g. $F(CH_2)_4CN \rightarrow F(CH_2)_3COOH + HCN$
toxic — toxic

It is clear that nitriles are not hydrolyzed appreciably *in vivo* since the toxic odd members would give rise to non-toxic acids, whereas this is plainly not the case; again, reduction cannot be the main metabolic reaction, since the toxic odd nitriles would form the corresponding odd amines, and these have been shown to be non-toxic (p. 148). Hence the most satisfactory way of explaining the high toxicity of the odd members is to accept the less plausible theory of C-CN rupture, resulting in the formation of toxic *even* fluoroacids and hydrogen cyanide; the process is further confirmed by the study of citric acid accumulation *in vivo* (p. 144). Thus the method has afforded independent evidence for the original suggestion of nitrile metabolism (p. 143).

A simpler approach can be used in the examination of certain types of linkages; the mere testing of one member can provide presumptive evidence in these cases. For example, in work with 'sesqui-fluoro-H', $FCH_2CH_2SCH_2CH_2SCH_2CH_2F$[80], the lack of toxicity and of any kind of fluoroacetate activity must surely imply that the thioether link in this compound is stable *in vivo* and that no enzyme system is capable of rupturing it to form fluoroethanol or fluoroacetic acid.

It is perhaps worth recording in some detail yet another example of the ω-fluorine atom acting as a tracer. Branched-chain fatty acids occur in synthetic fats, and consequently their metabolism is of some medical importance. What would be the effect of a branch methyl group on the breakdown of a fatty acid? An answer has been obtained[65] from an examination of the corresponding ω-fluoro compounds.

Considering the general formula $CH_3(CH_2)_nCHMe(CH_2)_m$-COOH, the following routes of degradation might be considered possible:

a) *m* odd:

$$CH_3(CH_2)_nCHMe(CH_2)_mCOOH \xrightarrow{\beta\text{-oxidation}} CH_3(CH_2)_nCHMeCH_2COOH \begin{cases} \nearrow CH_3(CH_2)_{n-1}COOH \\ ? \\ \searrow CH_3(CH_2)_nCOOH \end{cases}$$

b) *m* even:

$$CH_3(CH_2)_nCHMe(CH_2)_mCOOH \xrightarrow{\beta\text{-oxidation}} CH_3(CH_2)_nCHMeCOOH \begin{cases} \nearrow CH_3(CH_2)_{n-1}COOH \\ ? \\ \searrow CH_3(CH_2)_nCOOH \end{cases}$$

From this, it is clear that the effect of *m* is more important than that of *n*, since β-oxidation can occur after the branched methyl group has been passed. Moreover, it was considered likely that β-oxidation would occur up to the branched methyl group. Thus the problem appeared to be reduced to the metabolism of the α- and β-methyl derivatives. As indicated in the above scheme, for each of these there are two possible degradative routes.

The acids listed in Table XXXI were prepared by anodic coupling reactions[65].

TABLE XXXI

TOXICITY OF BRANCHED ω-FLUOROCARBOXYLIC ACIDS, $F(CH_2)_nCHMe(CH_2)_mCOOH$

Acid	LD_{50} *mg/kg*	LD_{50} *of corresponding unbranched acid, mg/kg*
a) *m* odd		
$F(CH_2)_5CHMeCH_2COOH$	118	0.64
$F(CH_2)_7CHMeCH_2COOH$	65	1.5
$F(CH_2)_8CHMeCH_2COOH$	2.4	57.5
b) *m* even		
$F(CH_2)_7CHMe(CH_2)_2COOH$	52	57.5
$F(CH_2)_8CHMe(CH_2)_8COOH$	2.7	5.7

When *m* is odd, the introduction of the methyl group causes a complete reversal of toxicological properties. This can only mean that a four-carbon fragment is being eliminated (possibly in two steps):

$$\underset{\text{non-toxic}}{F(CH_2)_5CHMeCH_2COOH} \rightarrow \underset{\text{non-toxic}}{F(CH_2)_4COOH}$$

On the other hand, when *m* is even, the presence of the methyl group has no appreciable effect on the overall toxicity pattern (within the limits of biological variation). Here the conclusion must be that in the process of β-oxidation, the methyl group is eliminated as part of a three-carbon fragment:

$$\underset{\text{toxic}}{F(CH_2)_8CHMe(CH_2)_8COOH} \rightarrow F(CH_2)_8CHMeCOOH \rightarrow \underset{\text{toxic}}{F(CH_2)_7COOH}$$

References p. 187

The fundamental difference in the two mechanisms can be seen in the case of the two isomeric methylundecanoic acids: the mere shift of the methyl group to the adjacent carbon atom results in one acid being over twenty times more toxic than the other. In summary, all compounds in which *n* is odd are non-toxic whereas those in which *n* is even are toxic, irrespective of *m* and hence of the total length of the carbon chain.

Applying these results to the original non-fluorinated branched acids, $CH_3(CH_2)_nCHMe(CH_2)_mCOOH$, it is reasonable to conclude that their breakdown occurs by similar pathways:

a) *m* odd:

$$\xrightarrow{\beta\text{-oxidation}} CH_3(CH_2)_nCHMeCH_2COOH \rightarrow CH_3(CH_2)_{n-1}COOH$$

b) *m* even:

$$\xrightarrow{\beta\text{-oxidation}} CH_3(CH_2)_nCHMeCOOH \rightarrow CH_3(CH_2)_{n-1}COOH$$

It is possible that these conclusions may be applied to fatty acids containing several non-adjacent methyl groups spaced on different carbon atoms along the chain, according to the following generalization (numbering from the COOH as carbon 1):

Position of CH_3	*CH_3 behaves as though it were*
any odd-numbered C	-CH_2- in the chain
any even-numbered C	not present

MISCELLANEOUS

Considerable work has been done on the effect of fluoroacetates and other monofluoro compounds in the control of micro-organisms, but no clearcut picture emerges as yet. It has been reported that the spread of some moulds (for example, *Physarella oblonga*, Morgan) can be reduced by fluoroacetate[1]; and organic materials subject to decomposition by micro-organisms are sterilized by the addition of small quantities of fluoroacetate esters[43]. Sodium fluoroacetate produced no pronounced beneficial effects in the control of poliomyelitis virus in mice[3, 41] or in monkeys[23]; similar remarks apply to the Eastern equine encephalomyelitis virus[94], the mumps virus[53], the pneumonia virus[53] and the influenza virus[2, 53] in mice. Sodium fluoroacetate administered one hour prior to infection affords some protection against vaccinia virus in rabbits[42]. Temporary inhibition of the growth of certain bacteria (for example, *Escherichia coli*) has been reported[51], but in general results are unpredictable[9, 10]; in some instances, fluoroacetate has been shown to lower the resistance of infected animals and hence to decrease the survival time.

Many fluoroacetates and long-chain ω-fluoro compounds have been screened against the mouse sarcoma 180 with no positive results[58]; some derivatives of 2-fluoroethyl urethane have also been examined as anticarcinogens[56]. Potter and Busch[73] have reported that tumour tissue (unlike normal tissue) in the rat does not show an accumulation of citrate after treatment with fluoroacetate, and hence cannot effectively metabolize fluoroacetate. Dietrich and Shapiro[20] have extended this observation to six different mouse neoplasms, but showed that in the seventh (the adenocarcinoma 755) there was citric acid accumulation accompanied by significant carcinostatic activity. Further work is necessary to clarify these findings. Finally, it may be mentioned that 2-fluoroethyl methanesulphonate, $FCH_2CH_2OSO_2CH_3$ was found[27] to have cytotoxic activity in experiments involving the transplanted Walker rat carcinoma.

REFERENCES

1. Abbott, C. E. (1945) The effects of DDT and of sodium monofluoracetate upon *Physarella oblonga*, Morgan. *Science, 102:* 71.
2. Ackermann, W. W. (1951) The relation of the Krebs cycle to viral synthesis. II. The effect of sodium fluoroacetate on the propagation of influenza virus in mice. *J. Exptl. Med., 93:* 635.
3. Ainslie, J. D. (1952) The growth curve of the Lansing strain of poliomyelitis virus in mice. The effect of sodium monofluoroacetate

and methionine sulfoxime on the early phase of growth of the virus. *J. Exptl. Med., 95:* 9.

4. Anon. (1955) *Medical manual of chemical warfare.* Fourth Ed., Her Majesty's Stationery Office, London, pp. 12–17 and 18–37.
5. Anon. '*Fluothane*', *a new inhalation anaesthetic.* Booklet published by Imperial Chemical Industries, Ltd., Pharmaceutical Division, Wilmslow, Cheshire, England.
6. BACON, J. C. (1947) Method of preparing fluoroacetamide. *U.S. Patent 2, 416, 607,* February 25, 1947.
7. BACQ, Z. M., FISCHER, P., Herve, A., LIÉBECQ, C., and LIÉBECQ-HUTTER, S. (1958) Sodium fluoroacetate as a protecting agent against irradiation. *Nature, 182:* 175.
8. BERGMANN, E. D., MOSES, P., and NEEMAN, M. (1957) Studies on organic fluorine compounds. VIII. N-Substituted fluoroacetamides as insecticides and rodenticides. *J. Sci. Food Agr., 8:* 400.
9. BERRY, L. J., MERRITT, P., and MITCHELL, R. B. (1954) The relation of the tricarboxylic acid cycle to bacterial infection. III. Comparison of survival time of mice infected with different pathogens and given Krebs cycle inhibitors and intermediates. *J. Infectious Diseases, 94:* 144.
10. BERRY, L. J., and MITCHELL, R. B. (1954) Some metabolic aspects of host-parasite interaction using the albino mouse and *Salmonella typhimurium. J. Infectious Diseases, 95:* 246.
11. CHAPMAN, C., and PHILLIPS, M. A. (1955) Fluoroacetamide as a rodenticide. *J. Sci. Food Agr., 6:* 231.
12. CHENOWETH, M. B. (1949) Monofluoroacetic acid and related compounds. *J. Pharmacol. Exptl. Therap., II, 97:* 383. *Pharmacol. Revs., 1:* 383.
13. DAVID, W. A. L. (1950) Sodium fluoroacetate as a systemic and contact insecticide. *Nature, 165:* 493.
14. DAVID, W. A. L., and GARDINER, B. O. C. (1951) Investigations on the systemic insecticidal action of sodium fluoroacetate and of three phosphorus compounds on *Aphis fabae* Scop. *Ann. Appl. Biol., 38:* 91.
15. DAVID, W. A. L., and GARDINER, B. O. C. (1953) The systemic insecticidal action of sodium fluoroacetate and of three phosphorus compounds on the eggs and larvae of *Pieris brassicae* L. *Ann. Appl. Biol., 40:* 403.

16. DAVID, W. A. L., and GARDINER, B. O. C. (1954) The systemic insecticidal action of certain compounds of fluorine and of phosphorus on *Phaedon cochleariae* Fab. *Ann. Appl. Biol., 41:* 261.
17. DAVID, W. A. L., and GARDINER, B. O. C. (1955) The aphicidal action of some systemic insecticides applied to seeds. *Ann. Appl. Biol., 43:* 594.
18. DAVID, W. A. L., GARDINER, B. O. C., CHAPMAN, C., and PHILLIPS, M. A. (1958) Fluoroacetamide as a systemic insecticide. *Nature, 181:* 1810.
19. DEONIER, C. C., JONES, H. A., and INCHO, H. H. (1946) Organic compounds effective against *Anopheles quadrimaculatus.* Laboratory tests. *J. Econ. Entomol., 39:* 459.
20. DIETRICH, L. S., and SHAPIRO, D. M. (1956) Fluoroacetate and fluorocitrate antagonism of tumor growth. Effect of these compounds on citrate metabolism in normal and neoplastic tissue. *Cancer Research, 16:* 585.
21. EMLEN, J. T., and STOKES, A. W. (1947) Effectiveness of various rodenticides on populations of brown rats in Baltimore, Maryland. *Am. J. Hyg., 45:* 254.
22. ESQUIBEL, A. J., KRANTZ, J. C., TRUITT, E. B., LING, A. S. C. and KURLAND, A. A. (1958) Hexafluorodiethyl ether (Indoklon): its use as a convulsant in psychiatric treatment. *J. Nervous Mental Disease, 126:* 530.
23. FRANCIS, T., BROWN, G. C., and KANDEL, A. (1954) Effect of fluoroacetate upon poliomyelitis in monkeys. *Proc. Soc. Exptl. Biol. Med., 85:* 83.
24. FRIED, J. F., ROSENTHAL, M. W., and SCHUBERT, J. (1956) Induced accumulation of citrate in therapy of experimental lead poisoning. *Proc. Soc. Exptl. Biol. Med., 92:* 331.
25. GRATCH, I., PURLIA, P. L., and MARTIN, M. L. (1949) Effect of sodium fluoroacetate (1080) in poisoned rats on plague diagnosis procedures. Preliminary report. *U.S. Public Health Repts., 64:* 339.
26. GRYSZKIEWICZ-TROCHIMOWSKI, E., SPORZYNSKI, A., and WNUK, J. (1947) Recherches sur les composés organiques fluorés dans la série aliphatique. II. Sur les dérivés des acides mono-, di et trifluoroacétiques. *Rec. trav. chim., 66:* 419.
27. HADDOW, A., and ROSS, W. C. J. (1956) Tumour growth-inhibitory alkyl sulphonates. *Nature, 177:* 995.

28. HARRISON, P. K. (1949) New compounds as insecticides against the sweet-potato weevil. *Bur. Entomol. and Plant Quarantine, E–770.*
29. HECHENBLEIKNER, I. (1948) 2-Butoxyethyl fluoroacetate. *U.S. Patent 2,456,586*, December 14, 1948.
30. HORSFALL, J. L. (1946) 2-Ethylhexyl fluoroacetate. *U.S. Patent 2,409,859*, October 22, 1946.
31. HUGHES, J. H. (1947) Deratization of surface vessels by means of 1080 (sodium fluoroacetate). *Pests, 15, No. 9:* 20, 22, 24.
32. HUGHES, J. H. (1947) Deratization of surface vessels by means of 1080 (sodium fluoroacetate). *U.S. Public Health Repts., 62:* 933.
33. HUGHES, J. H. (1950) 1080 (sodium fluoroacetate) poisoning of rats on ships. *U.S. Public Health Repts., 65:* 1021.
34. HURTIG, H., and PATTISON, F. L. M. (1950–1954) Unpublished work.
35. JENKINS, R. L., and KOEHLER, H. C. (1948) Making 1080 safe. A case study in the safe manufacture and distribution of a hazardous chemical. *Chem. Ind., 62:* 232.
36. JENSEN, R., TOBISKA, J. W., and WARD, J. C. (1948) Sodium fluoroacetate (compound 1080) poisoning in sheep. *Am. J. Vet. Research, 9:* 370.
37. JONES, E. S., MAEGRAITH, B. G., and GIBSON, Q. H. (1953) Pathological processes in disease. IV. Oxidations in the rat reticulocyte, a host cell of *Plasmodium berghei. Ann. Trop. Med. Parasitol., 47:* 431.
38. KALMBACH, E. R. (1945) 'Ten-eighty', a war-produced rodenticide. *Science, 102:* 232.
39. KAREL, L. (1948) The rodenticidal activity of fluoroacetphenylhydrazide ('Fanyline') and its oral toxicity to several species, with a note on the toxicity of 64 other candidate rodenticides. *J. Pharmacol. Exptl. Therap., 93:* 287.
40. KAREL, L., and WITTEN, B. (1951) Rodenticide comprising fluoroacetphenylhydrazide. *U.S. Patent 2,572,867*, October 30, 1951.
41. KIM, K. H. (1956) Effect of fluoride on the Lansing strain of poliomyelitis virus. *Virus (Japan), 6:* 59.
42. KIM, K. H. (1956) The effect of fluorides on vaccinia virus in rabbits. *Acta Schol. Med., Univ. Kioto, 33:* 145.
43. KLINGER, C. (1955) Verfahren zur Unterdrückung des Wachstums von Mikroorganismen. *Austrian Patent 185,502*, August 15, 1955.

44. KRANTZ, J. C., ESQUIBEL, A. J., TRUITT, E. B., LING, A. S. C., and KURLAND, A. A. (1958) Hexafluorodiethyl ether (Indoklon) – an inhalant convulsant. Its use in psychiatric treatment. *J. Am. Med. Assoc., 166:* 1555.
45. KRANTZ, J. C., PARK, C. S., TRUITT, E. B., and LING, A. S. C. (1958) Anesthesia. LVII. A further study of the anesthetic properties of 1,1,1-trifluoro-2,2-bromochloroethane (Fluothane). *Anesthesiology, 19:* 38.
46. KRANTZ, J. C., TRUITT, E. B., LING, A. S. C., and SPEERS, L. (1957) Anesthesia. LV. The pharmacologic response to hexafluorodiethyl ether. *J. Pharmacol. Exptl. Therap., 121:* 362.
47. LITZKA, G. (1936) Allgemeine biologische Wirkungen einer kernfluorierten Aminosäure (Fluortyrosin). *Naunyn-Schmiedeberg's Arch. exptl. Pathol. Pharmakol., 183:* 427.
48. LITZKA, G. (1936) Die antithyreotoxische Wirkung des Fluortyrosins. *Naunyn-Schmiedeberg's Arch. exptl. Pathol. Pharmakol., 183:* 436.
49. LU, G., LING, J. S. L., and KRANTZ, J. C. (1953) Anesthesia. XLI. The anesthetic properties of certain fluorinated hydrocarbons and ethers. *Anesthesiology, 14:* 466.
50. MACCHIAVELLO, A. (1946) Plague control with DDT and '1080'. Results achieved in a plague epidemic at Tumbes, Peru, 1945. *Am. J. Public Health, 36:* 842.
51. MAGER, J., GOLDBLUM-SINAI, J., and BLANK, I. (1955) Effect of fluoroacetic acid and allied fluoroanalogues on growth of *Escherichia coli*. I. Pattern of inhibition. *J. Bacteriol., 70:* 320.
52. MCCOMBIE, H., and SAUNDERS, B. C. (1946) Fluoroacetates and allied compounds. *Nature, 158:* 382.
53. MOGABGAB, W. J., and HORSFALL, F. L. (1952) Effect of sodium monofluoroacetate on the multiplication of influenza viruses, mumps virus, and pneumonia virus of mice (PVM). *J. Exptl. Med., 96:* 531.
54. OLAH, G., and PAVLATH, A. (1954) Synthesis of organic fluorine compounds. VI. Some derivatives of 2-fluoroethanol of insecticidal effect. *Acta Chim. Acad. Sci. Hung., 4:* 89.
55. OLAH, G., and PAVLATH, A. (1954) Synthesis of organic fluorine compounds. VII. Fluorinated aromatic insecticides containing sulphur. *Acta Chim. Acad. Sci. Hung., 4:* 111.

56. OLAH, G., PAVLATH, A., and NOSZKO, L. H. (1955) Synthesis and investigation of organic fluorine compounds. XIII. Derivatives of 2-fluoroethyl urethane. *Acta Chim. Acad. Sci. Hung., 7:* 443.

57. OTT, E., PILLER, G., and SCHMIDT, H. J. (1956) Über eine neue Synthese der 5-Fluor-*n*-valeriansäure und ihrer Ester und Glyceride und über deren Anwendung zur Zahnhärtung und Prophylaxe der Zahnkaries. *Helv. Chim. Acta, 39:* 682.

58. PATTISON, F. L. M. (1951–1957) Samples submitted to Dr. C. Chester Stock, Sloan-Kettering Institute, 410 East 68th St., New York, 21, N.Y.

59. PATTISON, F. L. M. (1957) Insecticides. *British Patent 767,638*, February 6, 1957 (to the Minister of National Defence, Ottawa, Canada).

60. PATTISON, F. L. M. (1957) From war to peace: toxic aliphatic fluorine compounds. *Chem. in Can., 9, No. 8:* 27.

61. PATTISON, F. L. M. (1959) Systemic insecticides. *Canadian Patent 572,653*, March 24, 1959.

62. PATTISON, F. L. M., FRASER, R. R., MIDDLETON, E. J., SCHNEIDER, J. C., and STOTHERS, J. B. (1956) Esters of fluoroacetic acid and of 2-fluoroethanol. *Can. J. Technol., 34:* 21.

63. PATTISON, F. L. M., HOWELL, W. C., and WOOLFORD, R. G. (1957) Toxic fluorine compounds. XIII. ω-Fluoroalkyl ethers. *Can. J. Chem., 35:* 141.

64. PATTISON, F. L. M., and WOOLFORD, R. G. (1957) (±)-18-Fluoro-10-methyloctadecanoic acid (Fluorotuberculostearic acid). *J. Am. Chem. Soc., 79:* 2306.

65. PATTISON, F. L. M., and WOOLFORD, R. G. (1957) Toxic fluorine compounds. XVI. Branched ω-fluorocarboxylic acids. *J. Am. Chem. Soc., 79:* 2308.

66. PAYNE, N. M. C. (1949) Insecticidal use of fluoroacetamide. *U.S. Patent 2, 469, 340*, May 3, 1949.

67. PETERS, R. A. (1952) Lethal synthesis. [Croonian lecture] *Proc. Roy. Soc. London, B, 139:* 143.

68. PETERS, R. A. (1957) Mechanism of the toxicity of the active constituent of *Dichapetalum cymosum* and related compounds. *Advances in Enzymology and Related Subjects of Biochemistry*, Vol. XVIII, Interscience Publishers, Inc., New York, p. 113.

69. PHILLIPS, M. A. (1954) The preparation of fluoroacetamide and sodium fluoroacetate. *Ind. Chemist, 30, No. 350:* 122.
70. PHILLIPS, M. A. (1955) The fluoroacetate series of pesticides. *World Crops, 7, No. 12:* 1.
71. PHILLIPS, M. A. (1957) Plant for the manufacture of fluoroacetamide. *Chemical Age*, April 20, 1957.
72. PHILLIPS, M. A., and WORDEN, A. N. (1956) Toxicity of fluoroacetamide. *Lancet, 271:* 731.
73. POTTER, V. R., and BUSCH, H. (1950) Citric acid content of normal and tumor tissues *in vivo* following injection of fluoroacetate. *Cancer Research, 10:* 353.
74. RAVENTÓS, J. (1956) The action of 'fluothane' – a new volatile anaesthetic. *Brit. J. Pharmacol., 11:* 394.
75. ROBBINS, B. H. (1946) Preliminary studies of the anesthetic activity of fluorinated hydrocarbons. *J. Pharmacol. Exptl. Therap., 86:* 197.
76. ROBINSON, W. B. (1950) Coyotes controlled by 1080. *Pest Control, 18, No. 6:* 12.
77. ROBINSON, W. B. (1953) Coyote control with compound 1080 stations in national forests. *J. Forestry, 51:* 880.
78. SARTORI, M. F. (1951) New developments in the chemistry of war gases. *Chem. Revs., 48:* 225.
79. SAUNDERS, B. C. (1952) Some aspects of the organic chemistry of fluorine and phosphorus. *School Sci. Rev., 33:* 320.
80. SAUNDERS, B. C., and STACEY, G. J. (1949) Toxic fluorine compounds containing the C-F link. Part IV. (a) 2-Fluoroethyl fluoroacetate and allied compounds. (b) 2,2′-Difluorodiethyl ethylene dithioglycol ether. *J. Chem. Soc., 1949:* 916.
81. SCALES, J. W. (1945) '1080' rat poison very effective but dangerous later. *Mississippi State Coll. Agr. Expt. Sta., Farm Research, 8, No. 12:* 8.
82. SCHRADER, G. (1945) The development of new insecticides. Presented by MUMFORD, S. A., and PERREN, E. A. in *British Intelligence Objectives Sub-committee, Report No. 714.*
83. SCHRADER, G., *et al.* (1946) Developments in methods and materials for the control of plant pests and diseases in Germany. Reported by MARTIN, H., and SHAW, H. in *British Intelligence Objectives Sub-committee, Report No. 1095.*

84. STRAUSS, B. S. (1956) The nature of the lesion in the succinate-requiring mutants of *Neurospora crassa:* interaction between carbohydrate and nitrogen metabolism. *J. Gen. Microbiol., 14:* 494.

85. SUCKLING, C. W. (1957) Some chemical and physical factors in the development of 'fluothane'. *Brit. J. Anaesthesiol., 29:* 466.

86. TAKEUCHI, T. (1954) Studies on organic fluorine compounds. Part III. On the contact action of organic fluorine compounds on the insect. *Ann. Rept. Takamine Lab., 6:* 135.

87. TEW, R. P. (1951) Pest control in Germany during the period 1939–1945. *British Intelligence Objectives Sub-committee Surveys, Report No. 32.*

88. TIETZE, E., SCHEPSS, W., and HENTRICH, W. (1930) Verfahren zum Schützen von Wolle und dergleichen gegen Mottenfrass. *German Patent 504,886,* August 9, 1930.

89. WALKER, G. L. (1957) Personal communication from Dr. E. E. GILBERT, General Chemical Division, Allied Chemical and Dye Corporation, Morristown, N.J., April 26, 1957.

90. WARD, J. C. (1946) Rodent control with 1080, ANTU and other war-developed toxic agents. *Am. J. Public Health, 36:* 1427.

91. WARD, J. C. (1956) How toxic are today's pesticides? *Pest Control, 24, No. 1:* 9, 12.

92. WARD, J. C. and SPENCER, D. A. (1947) Notes on the pharmacology of sodium fluoroacetate: compound 1080. *J. Am. Pharm. Assoc., 36:* 59.

93. WATANABE, S. (1955) 2-Chloroethyl fluoroacetate. *Japanese Patent 2300,* April 5, 1955.

94. WATANABE, T., HIGGINBOTHAM, R. D., and GEBHARDT, L. P. (1952) Effect of sodium monofluoroacetate on multiplication of Eastern equine encephalomyelitis virus. *Proc. Soc. Exptl. Biol. Med., 80:* 758.

6

Summary and final remarks

In the preceding pages, an attempt has been made to present a general survey of the preparations, properties and uses of toxic aliphatic fluorine compounds. As mentioned in the Preface, the emphasis has been more on balanced presentation than on exhaustive coverage of the literature. A general bibliography relevant to the subject is given in Appendix IV for more detailed reading.

It is perhaps convenient at this point to summarize the contents of the foregoing chapters:

Chapter 1 contains an *introductory account* of the subject.

Chapter 2 presents a general review of the *fluoroacetates*, including their occurrence, historical development, chemistry, physical and chemical properties, correlation between chemical structure and toxicity, toxicology, pharmacological aspects, biochemical aspects and mode of action, and finally medical aspects including diagnosis, therapy and post-mortem detection.

Chapter 3 reviews the present knowledge of *ω-fluorocarboxylic acids and derivatives*, including their occurrence, preparation, toxicology, β-oxidation as an explanation for the toxicological pattern, pharmacological aspects, and finally an account of miscellaneous derivatives.

Chapter 4 contains descriptions of *other ω-fluoro compounds* which possess toxic properties and of the uses in biochemistry to which these properties may be turned. Chemical series discussed include fluorinated hydrocarbons, oxygen-containing ω-fluoro compounds and ω-fluoro compounds containing nitrogen or sulphur.

Chapter 5 surveys *uses or potential uses* of toxic aliphatic fluorine compounds. It is a happy thought that what started as a cattle poison and candidate chemical warfare agent should have already proved so valuable to the pest exterminator, the biochemist and the pharmacologist, and should hold out such promise to the agriculturalist and the medical practitioner.

In general, the majority of the compounds which have been discussed are stable, colourless liquids, with odours characteristic of the non-fluorinated analogues. Their boiling points are often some 35° to 45° higher than those of the non-fluorinated compounds, some 25° to 35° lower than those of the corresponding chloro compounds, and some 40° to 50° lower than those of the corresponding bromo compounds (referred to atmospheric pressure). The outstanding stability of the C-F link results in chemical properties very similar to those of the non-fluorinated analogues; in fact, the fluorine resembles hydrogen more closely than it resembles the other halogens in regard to chemical properties. Indirectly, this resemblance results in the toxic properties of these fluorine compounds (p. 26).

The biochemical reaction common to all these compounds and responsible for their lethal action appears to be the conversion of 'active fluoroacetate' (fluoroacetyl-coenzyme A) to fluorocitric acid, which in turn blocks the tricarboxylic acid cycle by inhibiting competitively the enzyme aconitase; this blockage is thought to result in deprivation of energy leading to gross organic dysfunction and death. It follows that any compound which can give rise to fluoroacetyl-coenzyme A is likely to be toxic, and that such toxicity is likely to be proportional to the quantity of fluoroacetyl-coenzyme A which can be generated. This leads to the following general rules for predicting if a new compound is likely to be toxic or not:

(a) More than one fluorine atom on the same carbon atom usually results in a non-toxic compound.

There are of course exceptions to this rule; for example, perfluoroisobutylene, $(CF_3)_2C{=}CF_2$ and other polyfluoro olefins are toxic and even polytetrafluoroethylene (Teflon) at elevated temperatures is dangerous (p. 125).

(b) Toxicity is usually associated with a single fluorine atom situated in a terminal position in an aliphatic compound.

(c) Any compound which can give rise to fluoroacetyl-coenzyme A by some simple biochemical process is likely to be toxic.

Thus, rule (c) embraces fluoroacetic acid itself (by activation *in vivo*), and hence fluoroethanol (by oxidation), simple derivatives of these two compounds such as esters and amides (by hydrolysis and/or oxidation), toxic members of the ω-fluorocarboxylates (by β-oxidation), and the various classes of toxic compounds discussed in Chapter 4 (by metabolic removal of the functional groupings and then β-oxidation). Other subsidiary rules which permit a more exact interpretation of rule (c) are given on p. 158.

In spite of the fact that these compounds apparently operate by a common biochemical pathway, no satisfactory single antidote has yet been developed which is effective against both short- and long-chain members. Lacking any indications to the contrary, monoacetin therapy (see pp. 37, 55) is probably the procedure most likely to be successful in protecting against the effects of these poisons. However, as explained on p. 96, this treatment may be less effective in cases involving the long-chain members.

Although much has been achieved, much still remains to be done. Unquestionably, the outstanding need is for just such an antidote which is effective in all animal species and for all toxic members (fluoroacetates and long-chain compounds alike). When this has been found, the stock owners in South Africa can feel that the diverse researches over the past years have been justified in concrete form of real and lasting benefit; and medical practitioners can proceed with greater confidence in the treatment of victims. Further work is still required to vindicate some of the proposed uses and applications which for the most part must be considered as unproved. It is gratifying nevertheless that the preoccupation and emphasis has now shifted from toxicology and killing to biochemistry and healing; in short, that the ω-fluoro pendulum has made the sweep from war to peace. In this regard, however, it should be observed that the sentiments expressed in the quotations from Chaucer and Shakespeare at the start of the book represent hope, rather than achievement; a prayer, rather than a paean of victory.

Appendices

I. AN INTRODUCTION TO THE CHEMISTRY OF MONOFLUORO COMPOUNDS

Monofluorination

It has been stressed throughout this monograph that toxic aliphatic fluorine compounds are most likely to be encountered in Industry as *by-products* in manufacturing processes. Consequently, it is perhaps desirable to present a brief summary of preparative methods; this may serve as a guide in assessing possible hazards in any particular operation involving fluorine compounds. Literature references have not been included, because many appear elsewhere in the monograph and others are readily available from the reviews quoted in the General Bibliography (p. 210). In short, the purpose of this section is to provide a general introduction for those not familiar with the field.

The methods are classified according to the inorganic reagents involved in the formation of the C-F bond. The presence of functional groups in the molecule usually does not affect the reaction; thus, in many of the examples which follow, R may be aromatic or aliphatic and contain various substituents. To keep the review within reasonable bounds, experimental conditions have not been specified, because these can vary widely even for the same reaction; for example, in halogen exchanges involving potassium fluoride, satisfactory results have been obtained in glassware or in a high-pressure autoclave, with or without solvents, at temperatures ranging from 110° to 260°, under atmospheric or reduced pressure, with or without catalysts.

(a) *Fluorine*. The reaction of elementary fluorine with organic materials usually gives rise to a mixture of polyfluorinated products; but occasionally monofluoro compounds are formed which in at least two instances have been toxic. In the presence of other halogens, monofluorination by addition may occur.

$$CH_3COOH \text{ (as acid fluoride)} + F_2 \rightarrow FCH_2COOH + HF$$
$$CH_3CH_2CH_2COOH \text{ (as acid chloride)} + F_2 \rightarrow FCH_2CH_2CH_2COOH + HF$$
$$F_2 + Br_2 \rightarrow 2\ FBr;\ CH_2{=}CH_2 + FBr \rightarrow FCH_2CH_2Br$$

(b) *Hydrogen fluoride*. The equations which follow summarize some of the many monofluorinations brought about by hydrogen fluoride. From a preparative standpoint, probably the most promising general method involves the opening of epoxide rings.

$$RCH{=}CHR' + HF \rightarrow RCHFCH_2R'$$
$$RC{\equiv}CR' + HF \rightarrow RCF{=}CHR'$$
$$CH_2{=}C{=}O + HF \rightarrow CH_3COF$$
$$ROH + HF \rightarrow RF + H_2O$$
$$RCOCHN_2 + HF \rightarrow RCOCH_2F + N_2$$
$$RCH{-}CHR' \text{ (epoxide, bridged by O)} + HF \rightarrow RCH(OH)CHFR'$$
$$RCOCl + HF \rightarrow RCOF + HCl$$
$$Cl_3CNO_2 + HF \rightarrow Cl_3CF + HNO_2$$
$$(CH_2)_5\left<\begin{matrix}NNO\\|\\CO\end{matrix}\right. + HF \rightarrow F(CH_2)_5COOH + N_2$$
$$CO + CH_2O + HF \rightarrow FCH_2COOH$$

(c) *Metallic fluorides*. The following fluorides (listed by ascending atomic numbers) have been successfully employed in monofluorinations, usually by simple exchange reactions: NaF, KF, KHF_2, ZnF_2, AgF, SbF_3, Hg_2F_2, HgF_2, TlF and NH_4F. Some representative examples are listed below. In each instance, a specific fluoride is shown, but the same reaction can frequently be carried out using various different fluorides. In general, potassium fluoride is very much more effective than sodium fluoride; indeed, the former is probably the most satisfactory of all fluorides examined. Potassium bifluoride and zinc fluoride are useful only in certain cases (for example, the conversion of sulphonyl chlorides to sulphonyl fluorides, and of acyl chlorides to acyl fluorides, respectively). Argentous fluoride is of value in small-scale trial fluorinations of relatively stable halogen compounds, but the yields are often low. Antimony trifluoride, thallous fluoride and ammonium fluoride have not been examined extensively as monofluorinating agents. Conflicting reports have appeared regarding the merits of mercurous fluoride and mercuric fluoride, but the latter particularly appears to be an outstanding reagent.

$NaOSO_2OCH_2CH_2OH + NaF \rightarrow FCH_2CH_2OH + Na_2SO_4$
$RX + KF \rightarrow RF + KX\ (X = Cl, Br, I)$
$X(CH_2)_nX + 2\ KF \rightarrow F(CH_2)_nF + 2\ KX$
$X(CH_2)_nX + KF \rightarrow F(CH_2)_nX + KX$
$NH_4OSO_2OCH_2CONH_2 + KF \rightarrow FCH_2CONH_2 + KNH_4SO_4$
$ROSO_2OK + KF \rightarrow RF + K_2SO_4$
$ROSO_2R' + KF \rightarrow RF + R'SO_2OK$
$RSO_2Cl + KHF_2 \rightarrow RSO_2F + KCl + HF$
$ClCH_2CH_2OH + KHF_2 \rightarrow FCH_2CH_2OH + KCl + HF$
$2\ RCOCl + ZnF_2 \rightarrow 2\ RCOF + ZnCl_2$
$2\ ROSOCl + ZnF_2 \rightarrow 2\ ROSOF + ZnCl_2;\ ROSOF \rightarrow RF + SO_2$
$Br(CH_2)_nCOOR + 2\ AgF \rightarrow F(CH_2)_nCOOR + AgF.AgBr$
$CBr_4 + 2\ AgF \rightarrow CBr_3F + AgF.AgBr$
$3\ ClCH_2CONH_2 + SbF_3 \rightarrow 3\ FCH_2CONH_2 + SbCl_3$
$3\ CHBr_3 + SbF_3\ (Br_2) \rightarrow 3\ CHBr_2F + SbBr_3$
$2\ RX + Hg_2F_2 \rightarrow 2\ RF + Hg_2X_2$
$2\ ICH_2I + Hg_2F_2 \rightarrow 2\ FCH_2I + Hg_2I_2$
$2\ RX + HgF_2 \rightarrow 2\ RF + HgX_2$
$2\ BrCH_2CH_2COCH_3 + HgF_2 \rightarrow 2\ FCH_2CH_2COCH_3 + HgBr_2$
$RX + TlF \rightarrow RF + TlX$
$ROCOCl + TlF \rightarrow ROCOF + TlCl;\ ROCOF \rightarrow RF + CO_2$
$ROSO_2R' + NH_4F \rightarrow RF + R'SO_2ONH_4$
$ClCH_2CH_2OH + NH_4F \rightarrow FCH_2CH_2OH + NH_4Cl$

(d) *Miscellaneous preparations.* A few reactions have been reported which cannot be classified in the above three sections. Four examples follow.

$R_4NF \rightarrow RF + R_3N$
$R_4NBF_4 \rightarrow RF + BF_3 + R_3N$
$(C_6H_5)_3COH + CH_3COF \rightarrow (C_6H_5)_3CF + CH_3COOH$
$ROH + BrCOF \rightarrow ROCOF + HBr; ROCOF \rightarrow RF + CO_2$

Reactions

Monofluoro compounds, once formed, are usually very stable; moreover, the strong C-F bond allows the functional groups of monofluoro compounds to undergo a wide variety of reactions without simultaneous loss of fluorine. Thus, in the fluoroacetate series, the two compounds methyl fluoroacetate and 2-fluoroethanol can be used as the starting materials in the preparation of the many known members, by such reactions as oxidation, reduction, hydrolysis, substitution, dehydration, etc. (see pp. 22 and 24). Similar reactions have been carried out during the preparation of the long-chain compounds; this aspect of the work has been described at various places in the book, but for convenience, some of the more general methods may be mentioned, to illustrate the fact that the C-F bond can survive under a wide variety of experimental conditions.

(a) *Halide substitution reactions*

$F(CH_2)_nX + KCN \rightarrow F(CH_2)_nCN + KX$
$F(CH_2)_nX + AgNO_2 \rightarrow F(CH_2)_nNO_2 + AgX$
$F(CH_2)_nX + AgNO_3 \rightarrow F(CH_2)_nONO_2 + AgX$
$F(CH_2)_nX + KSCN \rightarrow F(CH_2)_nSCN + KX$
$F(CH_2)_nX + NaC{\equiv}CH \rightarrow F(CH_2)_nC{\equiv}CH + NaX$
$F(CH_2)_nX + AgOSO_2R \rightarrow F(CH_2)_nOSO_2R + AgX$
$F(CH_2)_nX + NaCH(COOEt)_2 \rightarrow F(CH_2)_nCH(COOEt)_2 + NaX$

(b) *Organometallic reactions*

$F(CH_2)_nX + Mg \rightarrow F(CH_2)_nMgX \rightarrow$ acids, esters, ketones, alcohols, aldehydes, etc.

$F(CH_2)_nX + Li \rightarrow F(CH_2)_nLi \rightarrow$ ketones, alcohols, etc.

(c) *Anodic coupling reactions*

$2\,F(CH_2)_nCOOH \rightarrow F(CH_2)_{2n}F + 2\,CO_2 + H_2$
$F(CH_2)_nCOOH + HOOC(CH_2)_mX \rightarrow F(CH_2)_{n+m}X + 2\,CO_2 + H_2$

(d) *Cyanoethylation*

$$F(CH_2)_nOH + CH_2{=}CHCN \rightarrow F(CH_2)_nOCH_2CH_2CN$$

(e) *Diels-Alder reactions*

$$\begin{array}{l} \text{cyclopentadiene } (CH{=}CH{-}CH_2{-}CH{=}CH \text{ ring}) + \begin{array}{c} CHCH_2F \\ \| \\ CHCOOR \end{array} \rightarrow \text{bicyclic adduct with } \begin{array}{c} CHCH_2F \\ | \\ CHCOOR \end{array} \end{array}$$

(f) *Lithium aluminium hydride reductions*

$$F(CH_2)_nCOOH \rightarrow F(CH_2)_nCH_2OH$$
$$F(CH_2)_nCOOR \rightarrow F(CH_2)_nCH_2OH$$
$$F(CH_2)_nCN \rightarrow F(CH_2)_nCH_2NH_2$$
$$F(CH_2)_nNO_2 \rightarrow F(CH_2)_nNH_2$$
$$F(CH_2)_nSCN \rightarrow F(CH_2)_nSH$$

II. SOME REPRESENTATIVE PREPARATIVE PROCEDURES

(a) 6-Fluorohexanol* (total halogen exchange)

A one-litre, three-necked, round-bottom flask (Note 1) is fitted with a thermometer, a precision-bore mechanical stirrer (Hershberg type, *Org. Syntheses*, Coll. Vol. *II* (1943) 117), and a reflux condenser, the upper end of which is protected by a calcium chloride tube. In the flask are placed 103 g of 6-chlorohexanol (Note 2), 125 g of anhydrous potassium fluoride (Note 3) and 500 g of diethylene glycol (Note 4). The mixture is stirred vigorously while being heated at 125° for 12–14 hours, and then cooled and poured into an equal volume of water. The crude product is isolated by thorough extraction with ether. After drying over calcium sulphate ('Drierite'), the extract is concentrated and the residue fractionated under reduced pressure. After a very small fore-run of unsaturated material, the yield of product boiling at 85–86°/14 mm is 59 g (65% of the theoretical amount) (Notes 5 and 6).

---

* Pattison, Howell, McNamara, Schneider and Walker, *J. Org. Chem.*, *21* (1956) 739.

Notes

1. The flask becomes slightly etched after several reactions, but may nevertheless be used indefinitely.

2. 6-Chlorohexanol (hexamethylene chlorohydrin) is prepared from hexamethylene glycol essentially by the method of Campbell and Sommers (*Org. Syntheses*, Coll. Vol. *III* (1955) 446). The yield is raised to 60–65% by packing the extraction flask with Berl saddles up to the level of the aqueous phase.

3. Potassium fluoride is dried at 160° for 24 hours, ground finely in a mortar or a ball mill, and then dried again; before use, it is ground once again while still hot and then stored in the oven at 160° for an additional 48 hours.

4. Commercial diethylene glycol is purified by distillation to constant boiling point and refractive index.

5. Other physical constants: n_D^{25} 1.4141; d_4^{20} 0.975.

6. Using the procedure described, the following ω-fluoro-alcohols have been prepared from the corresponding ω-chloro-alcohols:

	b.p.	n_D^{25}	d_4^{20}	*Yield %*
7-Fluoroheptanol	98–99°/12 mm	1.4197	0.956	62
8-Fluoro-octanol	106–107°/10 mm	1.4248	0.945	69
9-Fluorononanol	125–126°/15 mm	1.4279	0.928	66
10-Fluorodecanol	136–137°/15 mm	1.4322	0.919	64

(b) 6-Fluorohexyl chloride* (partial halogen exchange)

A one-litre, three-necked, round-bottom flask (Note 1) is equipped with a thermometer, a precision-bore mechanical stirrer (Hershberg type, *Org. Syntheses*, Coll. Vol. *II* (1943) 117), and a 25 cm Vigreux column fitted for vacuum distillation. In the flask are placed 168 g of 1,6-dichlorohexane (Note 2), 96 g of anhydrous potassium fluoride (Note 3) and 400 g of diethylene glycol (Note 4). The mixture is heated to 125° with constant vigorous stirring, and the pressure of the system is slowly reduced until a *slow*,

* Pattison and Howell, *J. Org. Chem.*, *21* (1956) 748.

steady rate of distillation is obtained (Note 5). Towards the end of the reaction, the pressure is very gradually reduced further, so as to maintain the constant rate of distillation. Finally, the pressure is held at 20 mm, and the temperature is slowly raised to 150° and maintained there until distillation ceases (Note 6). The distillate is washed with 10% sodium carbonate solution, and dried over anhydrous calcium chloride. The product is fractionated at atmospheric pressure to remove a small fore-run of 1,6-difluorohexane, b.p. 128–131°. The pressure is then reduced, and 65–72 g of 6-fluorohexyl chloride is collected, boiling at 61.5–62°/15 mm (Note 7). Further distillation provides 22–25 g of unchanged 1,6-dichlorohexane, boiling at 82–83°/11 mm. The yield of 6-fluorohexyl chloride based on reacted 1,6-dichlorohexane is 50–56% of the theoretical amount (Note 8).

Notes

1. The flask becomes slightly etched after several reactions, but may nevertheless be used indefinitely.
2. 1,6-Dichlorohexane may be prepared from hexamethylene glycol by reaction with thionyl chloride and pyridine.
3. Potassium fluoride is dried and ground finely as described in the previous procedure (Note 3).
4. Commercial diethylene glycol is purified by distillation to constant boiling point and refractive index.
5. A pressure of 90–100 mm is satisfactory. To obtain good yields, it is advisable to adjust the distillation rate to about one drop per two seconds throughout the reaction. If this rate is increased, more 1,6-dichlorohexane distils unchanged, and if it is reduced, more 1,6-difluorohexane is formed.
6. The reaction is complete in 4 to 5 hours.
7. Other physical constants: n_D^{25} 1.4168; d_4^{20} 1.015.
8. By essentially the same procedure, the following compounds have been prepared: 3-fluoropropyl bromide, 4-fluorobutyl chloride, 4-fluorobutyl bromide, 5-fluoroamyl chloride, 5-fluoroamyl bromide, 6-fluorohexyl bromide and 7-fluoroheptyl chloride.

(c) 1-Fluorohexane (*n*-hexyl fluoride)* (sulphonate cleavage)

1-Hexanol (15.3 g, 0.15 mole) and pyridine (30 g, 0.38 mole) are stirred at $-15°$, and to the mixture methanesulphonyl chloride (17.1 g, 0.15 mole) is slowly added. The mixture is stirred for a further 4 hours and diluted with water and 6*N* hydrochloric acid until there is no pyridine odour. The aqueous layer is extracted with methylene chloride, and the combined organic layer and extracts are washed with water. These are then transferred to a three-necked flask fitted with a thermometer, a precision-bore stirrer, and a still head carrying a condenser and ice-cooled receiver. The water-methylene chloride azeotrope is distilled, and any residual low-boiling material is removed under aspirator pressure. Diethylene glycol (250 g) and anhydrous potassium fluoride (17.4 g, 0.3 mole) are added, the temperature is raised to 100°, and the distillate is collected over $1\frac{1}{2}$ hours at a pressure of 300 mm. Fractionation of this yields 1-fluorohexane (8.8 g, 54% of the theoretical amount), b.p. 91–92°, n_D^{25} 1.3732.

(d) 1-Fluoro-2-heptanone** (diazomethyl ketone method)

Hexanoyl chloride (8.0 g, 0.06 mole) is added to a well-stirred ethereal solution of diazomethane (12.6 g, 0.3 mole), cooled in an ice-bath. Stirring is continued for 2 hours, and the ether is then removed. The residue (12 g) in ether (40 ml) is slowly added to liquid anhydrous hydrogen fluoride (5 g) in a polyethylene flask immersed in a Dry-Ice-acetone bath. The mixture is allowed to warm slowly to room temperature overnight, and then is poured over anhydrous potassium fluoride (20 g). The liquid is decanted, the solid cake is washed with dry ether, and the combined ethereal solutions are dried over anhydrous potassium fluoride. After

---

* PATTISON and MILLINGTON, *Can. J. Chem.*, *34* (1956) 757. A similar procedure has been described more recently by TITOV, VEREMEEV, SMIRNOV and SHAPILOV, *Doklady Akad. Nauk SSSR, Chemistry Section*, *113* (1957) 231 (Engl. translation); U.S.S.R. Patent 109,660, February 25, 1958. *p*-Toluenesulphonates are more effective than methanesulphonates, according to BERGMANN and SHAHAK, *Chem. & Ind.*, (1958) 157.

** FRASER, MILLINGTON and PATTISON, *J. Am. Chem. Soc.*, *79* (1957) 1959.

removal of the ether, the residue on fractionation gives 1-fluoro-2-heptanone (3.0 g, 38 % of the theoretical amount) of b.p. 54 °/13 mm and n_D^{25} 1.4048.

(e) Preparation of simple fluoroacetates from Compound 1080

*Fluoroacetic acid**. To technical sodium fluoroacetate (Compound 1080, containing 90% sodium fluoroacetate) (136 g) is added a mixture of 96% sulphuric acid (318 g) and 30% fuming sulphuric acid (232 g). After thorough mixing, the mixture is distilled under reduced pressure, yielding crude fluoroacetic acid, b.p. 75–80 °/11–14 mm. The pure crystalline acid (99.1 g, 93.5% of the theoretical yield) is obtained by distillation at atmospheric pressure, b.p. 167–168 °, m.p. 31–32 °.

*Ethyl fluoroacetate***. Technical sodium fluoroacetate (Compound 1080, containing 90 % sodium fluoroacetate) (60 g, 0.54 mole) and diethyl sulphate (100 g, 0.65 mole) are mixed and distilled. The distillate (b.p. 115–120 °) is dried over calcium sulphate and redistilled. Ethyl fluoroacetate thus is obtained as a colourless, pleasant-smelling liquid, b.p. 115–117 °, n_D^{25} 1.3750. Yield: 51 g, 90% of the theoretical amount.

Fluoroacetyl chloride°. Technical sodium fluoroacetate (Compound 1080, containing 90% sodium fluoroacetate) (167 g, 1.5 moles), previously dried for three days over phosphorus pentoxide, and phthalyl chloride (375 g, 1.68 moles) are mixed thoroughly and gently heated in a flask fitted for distillation. All the distillate of boiling point up to 90 ° is collected in a flask protected by a calcium chloride tube. The crude product is redistilled, yielding fluoroacetyl chloride (138 g, 95% of the theoretical amount) of b.p. 70–71 ° and n_D^{25} 1.3820.

---

* PATTISON, STOTHERS and WOOLFORD, *J. Am. Chem. Soc.*, *78* (1956) 2255.
** PATTISON, HUNT and STOTHERS, *J. Org. Chem.*, *21* (1956) 883.
° PATTISON, FRASER, MIDDLETON, SCHNEIDER and STOTHERS, *Can. J. Technol.*, *34* (1956) 21.

III. FIRST AID AND HOSPITAL TREATMENT

Lacking specific information to the contrary, the following recommendations may be considered as general for fluoroacetates and for all compounds which owe their toxicity to the ultimate formation of fluoroacetate. The instructions may be augmented by symptomatic treatment, at the discretion of the attending physician. Common-sense procedures, such as removal of contaminated clothing and washing of the skin, have not been included.

(a) FIRST AID *(N.B. Immediate treatment is of the utmost urgency)*

1. If the poison was swallowed and the patient is conscious and not convulsing, induce vomiting immediately by either of the following methods:

(a) Drinking 2 to 4 oz. of a *strong* solution of table salt in water (*i.e.* a solution containing as much table salt (sodium chloride) as will readily dissolve).

(b) Stimulation of the back of the throat with a spoon or padded stick (such as a wooden pencil with an erasing rubber attached). A finger may be used, but only as a last resort, since there is danger of its being bitten.

2. CALL A PHYSICIAN!

3. When available, monoacetin (glycerol monoacetate, glyceryl monoacetate) of any purity (Technical or better) (100 c.c. in 500 c.c. of water, *i.e.* approximately 3 oz. in a pint of water) may be drunk. Although the taste is unpleasant and further vomiting may occur, such treatment can do no harm but may do considerable good.

N.B. NEVER force fluids by mouth to unconscious or convulsing persons!

If a long delay is unavoidable in the arrival of the physician, a second identical dose of monoacetin may be taken after about one hour.

4. Keep the patient warm and quiet.

(b) DEFINITIVE TREATMENT

(To the physician: Hospitalization should be prompt; immediate treatment is of the utmost urgency.)

On the basis of a few human poisonings, the following symptoms and signs may be expected:

1. *Epileptiform convulsions* at any time from 30 minutes to 48 hours after ingestion of the poison.

They may be controlled by intravenous (or intramuscular) barbiturates by the usual procedure for severe convulsive states. They are *not* the usual cause of death.

2. *Cardiac irregularities.* Variation in the rate and rhythm may become very great. Usually there is alternation, progressing to failure of every other beat to be detectable at the radial pulse (pulsus alternans). Ventricular extrasystoles occur with increasing frequency as poisoning progresses, and death is usually produced by ventricular fibrillation or cardiac arrest.

Treatment

On the basis of animal (monkey) experiments (*J. Pharmacol. Exptl. Therap., 102* (1951) 31), it is to be anticipated that the following treatment may be beneficial. Monoacetin (glycerol monoacetate, glyceryl monoacetate) in large doses by intramuscular injection is a specific antagonist to fluoroacetate. The doses which might be employed are 0.1 to 0.5 c.c. per kg (2.2 lb.) body weight, *i.e.* 6 to 30 c.c. for an 11 stone (150 lb.) man. Some pain and oedema may be expected from such treatment. However, the toxicity of monoacetin is very low, and such slight respiratory stimulation, vasodilation, and sedation as may occur need not cause alarm.

The dosage should be repeated after about 30 minutes and whenever the patient's condition suggests progression of the poisoning. Continuous observation *by the physician* may be required for 48 hours because deterioration of the patient may occur rapidly and unexpectedly.

Notes

(1) If monoacetin is unavailable, acetamide dissolved in physiological saline may be used (*Biochem. J., 63* (1956) 182) by the

same procedure in approximately the same dosage; however, the historical background for the use of this material is less extensive.

(2) If the poison was swallowed, immediate and thorough gastric lavage should be of great value.

(3) Oxygen and artificial respiration may be applied if required.

MAYNARD B. CHENOWETH, M.D.,
Director of Pharmacological Research,
Biochemical Research Laboratory,
The Dow Chemical Company,
Midland, Mich. (U.S.A.)

IV. GENERAL BIBLIOGRAPHY

The following brief list of articles and reviews relating to toxic aliphatic fluorine compounds is not exhaustive, but it contains a representative selection which the author can recommend for more detailed reading.

(a) Emphasis on Chemistry

BOCKEMÜLLER, W. (1936) *Organische Fluorverbindungen. Sammlung chemischer und chemisch-technischer Vorträge, No. 28*, Verlag von Ferdinand Enke, Stuttgart, Germany.

HASZELDINE, R. N., and SHARPE, A. G. (1951) *Fluorine and its compounds*, Methuen and Co. Ltd., London, England.

LOVELACE, A. M., POSTELNEK, W., and RAUSCH, D. A. (1958) *Aliphatic fluorine compounds. ACS Monograph No. 138*, Reinhold Publishing Corporation, New York, N.Y.

SARTORI, M. F. (1951) New developments in the chemistry of war gases. *Chem. Revs., 48:* 225.

SAUNDERS, B. C. (1953) The chemistry and toxicology of organic fluorine and phosphorus compounds. *Royal Institute of Chemistry Lectures, Monographs and Reports, No. 1.*

SAUNDERS, B. C. (1957) *Some aspects of the chemistry and toxic action of organic compounds containing phosphorus and fluorine*, Cambridge University Press, England.

SCHIEMANN, G. (1951) *Die organischen Fluorverbindungen in ihrer Be-*

deutung für die Technik. Technische Fortschrittsberichte Band 52, Verlag von Dr. Dietrich Steinkopff, Darmstadt, Germany.

SCHRADER, G. (1951) *Die Entwicklung neuer Insektizide auf Grundlage organischer Fluor- und Phosphor-Verbindungen*, Verlag Chemie, G.m.b.H., Weinheim, Germany.

SIMONS, J. H. (1950 and 1954) *Fluorine chemistry*, Vol. I and II, Academic Press Inc., New York.

(b) Emphasis on Biochemistry, Pharmacology and Toxicology

BENSLEY, E. H., and JORON, G. E. (1958) *Handbook of treatment of acute poisoning*, Second Ed., E.S. Livingstone, Ltd., Edinburgh and London.

BREDEMANN, G. (1956) *Biochemie und Physiologie des Fluors und der industriellen Fluor-Rauchschäden*, Second Ed., Akademie-Verlag, Berlin, Germany.

CHENOWETH, M. B. (1949) Monofluoroacetic acid and related compounds. *J. Pharmacol. Exptl. Therap., II, 97:* 383. *Pharmacol. Revs., 1:* 383.

HARRISSON, J. W. E., AMBRUS, J. L., and AMBRUS, C. M. (1952) Fluoroacetate (1080) poisoning. *Ind. Med. and Surg., 21:* 440.

LARNER, J. (1950) Toxicological and metabolic effects of fluorine-containing compounds. *Ind. Med. and Surg., 19:* 535.

PATTISON, F. L. M. (1957) From war to peace: toxic aliphatic fluorine compounds. *Chem. in Can., 9, No. 8:* 27.

PETERS, R. A. (1952) Lethal synthesis [Croonian lecture]. *Proc. Roy. Soc., B, 139:* 143.

PETERS, R. A. (1954) Biochemical light upon an ancient poison: a lethal synthesis. *Endeavour, 13:* 147.

PETERS, R. A. (1955) Biochemistry of some toxic agents. II. Some recent work in the field of fluoroacetate compounds. [Dohme lectures.] *Bull. Johns Hopkins Hosp., 97:* 21.

PETERS, R. A. (1957) Mechanism of the toxicity of the active constituent of *Dichapetalum cymosum* and related compounds. *Advances in Enzymology and Related Subjects of Biochemistry*, Vol. XVIII, Interscience Publishers, Inc., New York, p. 113.

SMITH, F. A., and COX, G. C. (1952) A bibliography of the literature (1936–1952) on the pharmacology and toxicology of fluorine and its compounds, including effects on bone and teeth. [2449 entries.]

*Report** *issued by the University of Rochester*, Rochester, N.Y. (September 26, 1952).

(c) Emphasis on analysis**

BREDEMANN, G. (1956) *Biochemie und Physiologie des Fluors und der industriellen Fluor-Rauchschäden*, Second Ed., Akademie-Verlag, Berlin, Germany.

ELVING, P. J., HORTON, C. A., and WILLARD, H. H. (1954) Analytical chemistry of fluorine and fluorine-containing compounds. In *Fluorine Chemistry* (J. H. SIMONS, Editor), Vol. II, Academic Press Inc., New York, p. 51.

MACDONALD, A. M. G. (1959) Analysis of organic compounds containing fluorine. In *Comprehensive Analytical Chemistry* (C. L. WILSON and D. W. WILSON, Editors), Vol. IB, Elsevier Publishing Company, Amsterdam.

MCKENNA, F. E. (1951) Methods of fluorine and fluoride analysis. *Nucleonics*, McGraw-Hill Publishing Company, Inc., New York, *8, No. 6:* 24; *9, No. 1:* 40; *9, No. 2:* 51.

---

* An earlier report by F. A. SMITH (January 29, 1951), although containing fewer entries (1393), provides a brief summary of each.

** Several microanalytical laboratories in the United States undertake the analysis of fluorine in organic compounds; 3 to 10 mg of sample are required. The author recommends the Schwarzkopf Microanalytical Laboratory, 56–19 37th Avenue, Woodside 77, N.Y.

Index